Symbiosis Engine

The Chemistry of Cooperation in Nature

Team Trex

Published by NobleTrex Press

For permissions and other inquiries, write to:

P.O. Box 3132, Framingham, MA 01701, USA

Contents

Introduction

Take a deep breath. Hold it for a moment, then let it go.

That simple, rhythmic act-one you perform roughly twenty thousand times a day-is a deception. You likely perceive it as your own body at work: your diaphragm contracting, your lungs expanding, your blood greedily snatching oxygen from the air. But if we zoom in past the tissues, past the cells, and down to the molecular furnaces burning that oxygen to keep you alive, we find a stranger.

Deep inside nearly every cell of your body are mitochondria. They have their own DNA, distinct from yours. They reproduce on their own timetables. They are, by every biological definition, a separate entity. Billions of years ago, they were free-living bacteria swimming in a primordial soup, until they were engulfed by a larger, hungrier microbe. But instead of being digested, a truce was called. A trade agreement was signed in chemistry: the captive bacterium would provide energy in exchange for protection. That ancient merger did not just survive; it conquered the planet. It allowed life to scale up from microscopic specks to blue whales, redwood trees, and you.

You are not an individual. You are a colony, a walking, talking boardroom of negotiations that have been running without interruption for eons.

This book is the story of those negotiations. It is an exploration of the *Symbiosis Engine*, the relentless chemical drive that pushes life toward cooperation. For more than a century, popular science has been dominated by a specific, somewhat brutal narrative: nature as a gladiator pit. We speak of "survival of the fittest" and "nature red in tooth and claw," phrases that conjure images of lions taking down gazelles or trees strangling one another for a patch of sunlight. While competition is undeniably a potent sculptor of evolution, it is only half the story. The other half-the structural integrity that holds the biosphere together-is cooperation.

But this is not a story about benevolence or altruism. Nature does not offer charity. When we peer into the mechanics of how different species live together, we do not find kindness; we find economics. We find a ruthless, highly efficient stock market where the currency is not gold or digital credits, but phosphorus, nitrogen, carbohydrates, and proteins. We find a world held together by the cold, hard calculus of cost and benefit, brokered through the most sophisticated language on Earth: chemistry.

Why does this matter to us, standing as we do on the paved surface of the twenty-first century? Because the Symbiosis Engine powers the systems that keep us alive. The crops that fill our plates rely on negotiations with soil bacteria to access nutrients. The oxygen we breathe is the byproduct of oceanic collaborations. Even our own health is a diplomatic situation; the human gut is home to trillions of microbes that regulate our immune systems, synthesize our vitamins, and, as we are beginning to discover, influence our moods and minds. When these alliances hold, we thrive. When the chemical conversation breaks down-whether in a bleaching coral reef or an inflamed human colon-the system collapses.

To understand how life actually works, we have to learn to speak its language. For decades, biology was largely a science of description: counting legs, measuring leaves, and cataloging species. But a revolution has taken place. We can now eavesdrop on the chemical chatter passing between species. We know that plants scream in specific molecular frequencies when attacked by caterpillars, summoning predatory wasps to their aid. We know that trees in a forest are not isolated pillars of wood but are physically connected by fungal internet cables, trading sugars for water in a vast, underground economy.

Our guide into this world begins with the fundamental realization championed by the late evolutionary theorist Lynn Margulis, who forcefully argued that the merger of distinct organisms is the primary engine of evolutionary innovation. While her ideas were once considered heretical, the evidence has become overwhelming. We will see that the "self" is a fluid concept. The boundaries we draw between organisms are convenient fictions for our own categorization, but biology respects no such borders.

As we strip away the layers of the macroscopic world, we uncover the "Molecules of Meaning." This is the invisible vocabulary of life. In the air and water, organisms are constantly releasing volatile compounds-perfumes, stinks, and tasteless drifts of gas-that serve as invitations, warnings, and contracts. We will examine the work of pioneers like Tom Eisner and Jerrold Meinwald, who helped establish the field of chemical ecology, teaching us that to a blind worm or a root tip, the world is not a landscape of shapes and colors, but a swirling topographic map of scents and gradients.

It is easy to romanticize these connections, to view the "Wood Wide Web" or the coral reef as a utopia of sharing. However, the chemistry of cooperation is strictly audited. In the darkness of the soil, a biological market is in session. Here, fungi and plant roots engage in trade that mimics the supply-and-demand curves of a human stock exchange. When phosphorus is scarce, the fungi raise their prices. If a plant partner stops paying its sugar dues, the fungus cuts off the supply line, suffocating the freeloader. This is the logic of the ledger. Evolution favors those who cooperate, yes, but only those who audit their partners.

This brings us to the inevitable shadow side of cooperation: the art of deception. Wherever there is a system of trust, there will be confidence artists. We will meet the orchids that mimic the scent of a female wasp to trick amorous males into pollination without offering a drop of nectar in return. We will look at the parasites that hack the chemical passwords of an ant colony, living like kings inside the fortress of their enemies. These cheaters are not merely villains; they are essential drivers of the system, forcing honest partners to develop more sophisticated encryption and policing methods. It is an evolutionary arms race, not of claws and teeth, but of passwords and keys.

The stakes of understanding this chemistry have never been higher. We are currently witnessing the breakdown of ancient partnerships on a planetary scale. The coral reefs, perhaps the most visually stunning example of symbiosis, are bleaching white as ocean temperatures rise. This bleaching is a chemical divorce: the heat disrupts the molecular conversation between the coral animal and its algal tenants, causing the host to expel the very partners it needs to survive. By understanding the specific molecular breakdown that occurs, scientists are racing to engineer heat-resistant partnerships, hoping to buy time for these underwater metropolises.

We will also turn our gaze inward. The medical community is undergoing a paradigm shift, moving away from the "war on germs" mentality toward a management of internal ecology. We are realizing that the human body is a distinct ecosystem, a "holobiont." The bacteria in our gut do not merely digest our food; they produce metabolites that enter our blood stream and cross the blood-brain barrier. We are learning that distinct chemical signals from our microbiome can alter our anxiety levels, our appetite, and perhaps even our personality. The implications for medicine are profound: the future of psychiatry and oncology may lie not in synthesized drugs, but in restoring the chemical balance of our inner wildlife.

Finally, we must look to the future. For most of human history, our relationship with nature has been extractive. We mine the soil, we harvest the seas, we clear the forests. But a new era of "designer alliances" is emerging. In laboratories around the world, synthetic biologists are rewriting the genetic code to create new mutualisms-crops that fertilize themselves through engineered bacterial partnerships, or microbial consortia designed to eat plastic waste. If we can master the grammar of the Symbiosis Engine, we can move from being exploiters of the biosphere to active participants in its repair.

This book is an invitation to change how you see the world. When you walk through a park, I want you to see not just static greenery, but a roaring conversation. I want you to understand that the clover under your foot is engaged in a high-stakes nitrogen trade with the bacteria in its roots. I want you to realize that the bee buzzing past is following a trail of volatile compounds specifically formulated by a flower to manipulate its nervous system. And I want you to feel the weight of your own breath, acknowledging the silent, ancient partners in your cells that make your very existence possible.

The view of life as a solitary struggle is an illusion. We are, all of us, woven together by a delicate, invisible web of chemical need. Let us examine the threads.

Chapter 1

The Cooperative Imperative

Evolution is often depicted as a gladiator match, but the machinery of life runs on a quieter, more potent fuel. We uncover the chemical pacts that bind organisms together, revealing how cooperation is not an altruistic anomaly but a ruthless necessity for survival found in the very architecture of our cells.

1.1 Beyond Red in Tooth and Claw

The most familiar picture of evolution is a contest: organisms locked in a struggle for food, space, mates, and light. It is not a false picture. Scarcity is real, and living things do kill and outcompete one another. Yet the slogan "red in tooth and claw" can quietly smuggle in a deeper assumption: that the principal unit of evolution is a solitary individual, self-contained, negotiating the world alone. When that assumption hardens into habit, cooperation starts to look like an exception that needs special pleading. The history written into cells suggests the opposite. Some of the greatest leaps in biological complexity did not come from the refinement of weapons, but from the forging of alliances so intimate that the partners became inseparable.

A useful way to feel the scale of this claim is to notice something about our own bodies. Every second, muscles contract and relax; nerves maintain voltage differences across membranes; enzymes fold and unfold; immune cells patrol; kidneys filter. None of it is possible without an incessant stream of energy, and that stream depends on organelles called mitochondria. They are often described as "powerhouses," which is not wrong, but it is also slightly too tidy. Powerhouses sound like buildings; mitochondria are populations. They are numerous (hundreds to thousands in a typical cell), dynamic (they fuse and divide), and chemically opinionated (they decide, through signals, whether a cell should persist or self-destruct). In a sense that matters for the origins of complex life, mitochondria are not merely parts of us. They are the descendants of once-free bacteria.

The idea that mitochondria came from bacteria was not always mainstream. It had to compete with a simpler story: that complex cells arose by gradually elaborating internal membranes and metabolic pathways within a single lineage, step by step, with no decisive moment of merger. The alternative story—that a cell became complex by *acquiring* another cell and keeping it—sounds at first like science fiction. It can also sound, unhelpfully, like an allegory of predation: a larger cell eats a smaller one, end of story. But the crucial point is that this meal did not end with digestion. It ended with a truce that became a constitution.

Lynn Margulis became the most influential champion of this view, calling it *symbiogenesis*: the origin of new forms through symbiosis. Her advocacy in the late 1960s and 1970s was as much a human drama as a scientific argument, marked by skeptical reviews and repeated rejections before the idea took root. Part of the resistance was psychological: evolution had become fluent in small changes. A permanent cellular merger felt too large, too abrupt, too close to "hopeful monsters." Part of the resistance was evidential: where was the proof that mitochondria were not simply inventions of the cell itself?

The proof accumulated in several converging lines, each persuasive on its own, overwhelming together. Mitochondria contain DNA, and it is not arranged like the chromosomes in a cell nucleus. It is small, often circular, and more similar to bacterial genomes than to eukaryotic nuclear genomes. Mitochondria also have their own ribosomes, and these ribosomes resemble bacterial ribosomes in important structural and functional ways. Mitochondria divide by fission, in the manner of bacteria, rather than being assembled from scratch each time a cell divides. Their inner membranes are the sites of electron transport chains and proton gradients, the hallmarks of bacterial energy metabolism. Even the double membrane around mitochondria—an outer membrane and a highly folded inner membrane—makes a kind of anatomical sense if an engulfed bacterium retained its own boundary while acquiring a second layer from its host.

None of this requires that the host was benevolent, or even that it intended cooperation. Nature does not sign contracts. The earliest stages could have been unstable, opportunistic, and violent. An ancestral host cell (likely an archaeal lineage, as molecular phylogenies now suggest) may have engulfed an aerobic bacterium and initially treated it as food. The bacterium may have resisted digestion or evaded it by chemical means. Or the host may have benefited from keeping the bacterium alive because the bacterium could detoxify oxygen or harvest energy efficiently in an oxygenated world. What matters is the selective pressure that followed: any host lineage that could keep the bacterium and extract reliable benefit from it would have had a dramatic advantage, and any bacterium that could persist inside a host while exporting useful chemistry would have gained a safe niche rich in substrates.

The atmosphere sets the stage. Oxygen is a paradoxical substance: essential for the energetically rich metabolism that powers animals and plants, yet chemically reactive enough to damage proteins, lipids, and DNA. When oxygen rose in Earth's oceans and air, it transformed the rules of survival. Aerobic respiration—using oxygen as a final electron acceptor—is astonishingly efficient compared to many anaerobic alternatives. A cell that could harness oxygen safely could, in principle, afford a larger genome, more regulation, more internal architecture. The trouble is that oxygen's efficiency comes with oxidative stress. It produces reactive oxygen species that can puncture membranes and scramble macromolecules. One way to manage that danger is to outsource it: to keep oxygen-handling chemistry localized in a compartment whose damage can be managed, repaired, or, in extremis, sacrificed.

That is an engineering description, but it captures the chemical logic of endosymbiosis. The bacterium that became the mitochondrion was an expert in aerobic metabolism. The host cell, by internalizing this expert, gained a portable power plant whose dangerous byproducts could be sequestered. Over evolutionary time, the relationship tightened. Genes moved from the symbiont to the host nucleus, a process called *endosymbiotic gene transfer*. The symbiont's genome shrank, shedding functions that the host could supply. The host evolved protein import machinery to send nuclear-encoded proteins back into the mitochondrion, making it dependent. Dependency is often spoken of as weakness, but here it is also a stabilizer: once both partners lose the ability to live independently, defection becomes difficult.

At the scale of a single cell, this sounds like an exchange of services. At the scale of evolution, it is a redefinition of what an individual is. The complex cell—the eukaryotic cell—is not merely a bacterium with extra features. It is a consortium, a chimera of lineages, a cooperative arrangement that became so habitual it now feels like a single thing. The word *holobiont* is sometimes used for an organism understood together with its symbiotic partners, especially in the context of microbiomes. It is a provocative word because it suggests that selection may act not only on a host genome but on a network of genomes that persist together. One need not insist that holobionts are always the primary units of selection to appreciate the less controversial point: biological individuality is often constructed. It is not given.

Mitochondria are the canonical example, but they are not the only one. Plants carry chloroplasts, the organelles that perform photosynthesis, and chloroplasts also bear the signature of bacterial ancestry. In that case the bacterial partner was a cyanobacterium, capable of harvesting light and fixing carbon dioxide into sugars. The same logic holds: internalization, gene transfer, shrinking of autonomy, and the gradual evolution of interlocking dependencies. These events did not merely add a gadget to a cell; they reshaped global ecology. Oxygen in the atmosphere, the greening of continents, the energy budgets that made large bodies possible—all are tied to symbiotic mergers.

The deepest challenge to the solitary view of evolution is not that cooperation exists, but that cooperation can be creative in a way competition cannot easily mimic. Competitive adaptation tends to refine within an existing design space. Symbiosis can splice design spaces together. A host that acquires a symbiont can suddenly access metabolic pathways, enzymes, and membrane architectures that took the symbiont billions of years to evolve. This is not magic; it is inheritance by association. But it can produce the evolutionary equivalent of technological transfer: new capabilities appearing not through incremental invention, but through acquisition and integration.

Integration, however, is not automatic. A permanent merger faces an immediate problem: conflict of interest. The host would benefit from symbionts that devote resources to host needs; the symbiont would benefit from replication, perhaps at the host's expense. If symbionts reproduce inside a host, selection among symbionts could favor "selfish" variants that replicate faster even if they harm the host. Something must align incentives. In eukaryotes, one important alignment mechanism is a bottleneck: mitochondria are typically inherited through a narrow channel (often maternal inheritance), which reduces competition among unrelated symbiont lineages within a host. Another is host control: the host regulates mitochondrial division, distribution, and turnover. Cells also use a quality-control process called *mitophagy* to remove damaged mitochondria, a policing mechanism in chemical clothing. The details vary among lineages, but the principle is general: cooperation at a higher level requires suppression of conflict at a lower level.

This is where chemistry becomes a language of governance. The host does not merely contain mitochondria; it monitors them. Mitochondria send signals when they are stressed, when their membranes lose integrity, when electron transport is faltering. Some of those signals are metabolites; others are changes in redox state; others are bursts of reactive oxygen species that, in small controlled doses, function less like damage and more like messages. The host responds by altering gene expression, rerouting metabolism, or initiating removal. In extreme cases, mitochondria contribute to apoptosis, programmed cell death, by releasing proteins that activate cellular self-destruction pathways. That seems grim, but it is also a form of community hygiene: a cell that cannot maintain its internal cooperation can be eliminated for the sake of the multicellular organism.

A historical anecdote can make this feel less abstract. When early microscopists began to catalog the contents of cells, they saw granules and threads whose significance was unclear. The structures we now call mitochondria were described in the nineteenth century under various names, and their role was mysterious. It took the twentieth century's biochemistry to connect them to respiration. Even then, the idea that they were once bacteria was not an inevitable conclusion. Margulis's persistence mattered because she treated disparate observations as clues to a single story: DNA where it should not be, ribosomes with bacterial character, membranes nested like a swallowed cell. Scientific progress here was not a straight line from data to theory; it involved a willingness to take the strange features of mitochondria at face value and to ask what kind of past would make them unsurprising.

The endosymbiotic origin of mitochondria is now strongly supported, but that does not mean every detail is settled. The precise bacterial lineage closest to the mitochondrial ancestor remains an active topic in phylogenomics. Different analyses, using different gene sets and models, place the ancestor in slightly different neighborhoods within the Alphaproteobacteria. That uncertainty is not a weakness of the basic claim. It is what one expects when trying to reconstruct events that occurred deep in time, in lineages that have since diversified and exchanged genes. The broad outline is stable: a bacterium became an organelle. The fine print is still being negotiated.

What does this do to the emotional tone of evolution? It does not make nature gentle. Symbioses can begin as exploitation and stabilize into mutual dependence. They can also break down. A partner can become a parasite if conditions change. Even mitochondria, indispensable as they are, carry their own hazards: mitochondrial dysfunction is implicated in aging and in many diseases. Cooperation is not a moral achievement; it is a dynamic arrangement under constraint. Yet recognizing the role of merger and alliance changes the kinds of questions that feel natural to ask. Instead of only asking how an organism defeats rivals, one also asks how it recruits partners, keeps them honest, integrates their chemistry, and survives the tensions of intimacy.

This perspective also makes the origin of complex life feel less like a smooth ascent and more like a series of negotiated incorporations. Consider what a eukaryotic cell is able to do that bacteria generally cannot. It can maintain a nucleus that protects and regulates a large genome. It can build a cytoskeleton that changes shape, engulfs particles, and organizes internal traffic. It can evolve intricate signaling pathways and, eventually, multicellularity with specialized tissues. These capabilities require energy and control. Energetically, mitochondria supply high-yield respiration, but the crucial twist may be not merely more energy, but *energy per gene* or energy per unit of regulatory complexity: enough spare capacity to run a cell with many moving parts. Control, meanwhile, emerges from compartmentalization and communication. Organelles allow chemical processes that would interfere with one another to be separated in space and linked by regulated transport. Cooperation is thus not only about sharing; it is about creating boundaries, channels, and rules.

The word "engulfment" can sound like a single dramatic instant: one cell swallows another, and the rest follows. In reality, the path from association to organelle likely involved intermediate stages that were neither stable nor identical across lineages. Some associations may have started on surfaces, with bacteria living attached to host membranes, trading metabolites across a narrow gap. Some may have involved repeated ingestion, where bacteria were often eaten but occasionally survived, and host variants that were less efficient at digestion accidentally provided a refuge. Some may have involved hosts that evolved to tolerate a bacterial presence because the benefits outweighed the costs in a particular environment. Evolution is not obliged to take the shortest route. It is only obliged to find routes that work.

It is also worth noticing how much of this story is about membranes. Membranes are not passive wrappers; they are active chemical interfaces. They determine what can enter and leave, which reactions can occur nearby, which gradients can be maintained. The hallmark of mitochondria is the inner membrane's ability to maintain a proton gradient—a difference in proton concentration and electrical charge across a membrane. That gradient is the immediate currency of ATP production. In bacteria, this happens across the cell membrane. In eukaryotes, it happens inside an organelle, fenced off from the rest of the cell. The merger that produced mitochondria was, among other things, a merger of membranes and the chemical worlds they separate.

A schematic can help anchor the idea that endosymbiosis is not mere swallowing but the birth of a shared architecture.

Endosymbiosis: From Engulfment to Organelle

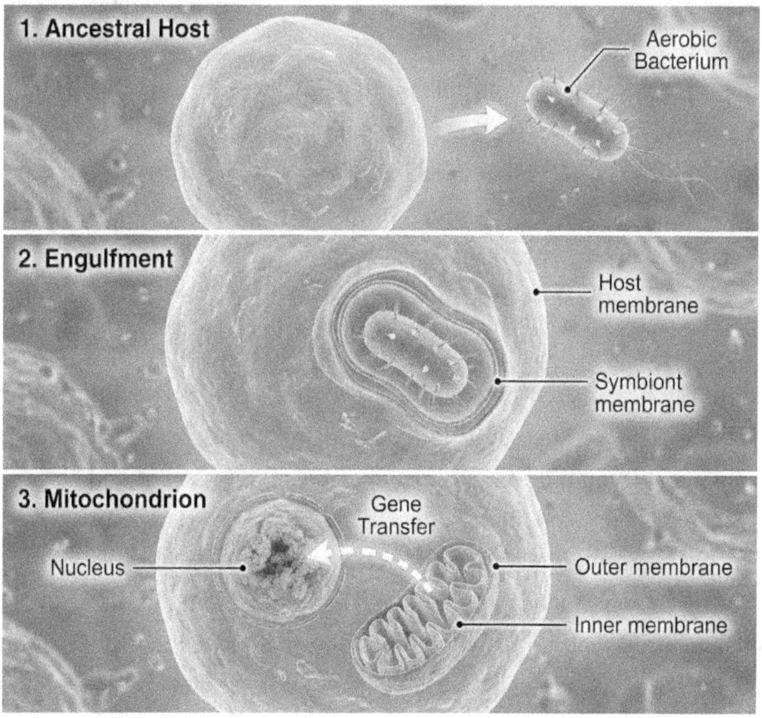

The diagram is necessarily simplified, but two features deserve attention because they capture the "merger" rather than the "meal." First, the persistence of two membranes. In many mitochondria, the inner membrane is the descendant of the bacterial membrane, and the outer membrane reflects the host-derived boundary that came with engulfment. Second, the arrow of gene transfer and integration. Modern mitochondria retain only a small genome compared with free-living bacteria. Much of what mitochondria do is executed by proteins encoded in the nucleus, manufactured in the cytosol, and imported into the organelle. That arrangement makes the mitochondrion less like a captured prisoner and more like a department within a larger institution: partially autonomous in day-to-day chemistry, but ultimately governed by a central genome.

Once one becomes comfortable with endosymbiosis, it becomes harder to tell a story of life that is only about individuals. Even within a single organism, the boundary of "self" is negotiated by membranes, signals, and shared metabolism. The mitochondrion is a reminder that cooperation is not merely a social strategy layered on top of biology. It is embedded in biology's architecture. A cell is not simply a bag of enzymes competing to run reactions. It is a managed economy of compartments, each with its own chemical constraints, kept in alignment by exchange and oversight.

That alignment has a moral that is easy to miss: cooperation can be brutally pragmatic. The host cell did not keep the bacterium out of kindness; it kept it because the chemistry worked. The bacterium did not surrender its autonomy out of loyalty; it surrendered because genes that once supported independence became redundant and were lost. The result is nevertheless extraordinary: a partnership so deep that neither partner, in anything like its original form, exists anymore. The energy that animates complex life still passes through that ancient bargain, proton by proton across a folded inner membrane, as if a long-ago alliance continues to be paid in the smallest possible coins.

1.2 The Spectrum of Intimacy

Once the idea of cellular mergers feels plausible, it becomes harder to keep the moral categories tidy. The comforting story goes like this: parasites are villains, mutualists are friends, and commensals are the harmless background crowd. Nature, inconveniently, does not respect these labels. The same pair of species can drift from helpful to harmful as temperature, nutrients, crowding, or predators change. Even within a single day, the balance can slide. "Symbiosis" simply means living together; it does not promise harmony. The more honest picture is a spectrum, with relationships moving along it as conditions alter the costs and benefits of staying close.

Start with the plain definitions, because they matter. In *mutualism*, both partners gain a net benefit. In *parasitism*, one gains while the other pays a net cost. *Commensalism* sits in the middle: one partner benefits while the other is, in principle, unaffected. These terms sound crisp, but they hide a measurement problem. Benefit and cost are not feelings. In biology they usually mean changes in survival and reproduction, or, more mechanistically, changes in energy, growth, or successful offspring. That makes the categories context-dependent by design. A relationship can be mutualistic at one life stage and parasitic at another, mutualistic under one diet and parasitic under another, mutualistic when partners are scarce and parasitic when they are abundant.

A simple human analogy helps without taking it too far. A roommate arrangement can be wonderful when both people value quiet, pay rent reliably, and share chores. Add sleep deprivation, financial stress, and a kitchen full of unwashed dishes, and the same arrangement becomes corrosive. The biology is less sentimental and more chemical, but the logic is similar: cooperation is often a bargain that only makes sense under certain constraints.

It is tempting to treat commensalism as a safe, neutral category: a barnacle on a whale, a bacterium riding along without consequence, a plant taking shade from another plant's canopy. Yet "no effect" is a strong claim. It is often a placeholder for "we have not measured the effect carefully enough." A microbe that seems neutral in a well-fed host might become costly during starvation by competing for scarce nutrients. A passenger can become a burden when the vehicle runs out of fuel. Conversely, a relationship that looks parasitic in a laboratory might soften into mutualism in a more realistic environment, where the "harm" is offset by protection against a third enemy.

The spectrum metaphor becomes especially vivid when the partners exchange things that can be counted: carbon, nitrogen, phosphorus, vitamins, defensive chemicals. Exchange makes intimacy legible. If one partner supplies something the other cannot make, dependence grows. If both supply something valuable, mutualism becomes possible. If one partner can take without paying, parasitism beckons. The spectrum is not only about who wins and loses, but about the *terms* of trade and how the terms shift.

A classic case where the terms can flip is the relationship between legumes (peas, beans, clover) and nitrogen-fixing bacteria called rhizobia. Nitrogen is abundant in the air, but atmospheric nitrogen gas is chemically inert enough that most organisms cannot use it directly. Plants need nitrogen in reduced forms to build proteins and nucleic acids. Rhizobia possess the enzyme machinery to turn atmospheric nitrogen into ammonia, a usable form. The partnership seems made for mutualism: bacteria get carbon compounds from the plant, the plant gets nitrogen.

But nitrogen fixation is expensive. The enzyme complex that does it is sensitive and energy-hungry. In a soil already rich in nitrogen fertilizer, the plant's need for bacterial nitrogen shrinks. Now the bargain changes. The plant still has to build and maintain root nodules, still has to supply carbon, still has to manage oxygen levels inside the nodule so the bacterial enzymes are not damaged. Under those conditions, rhizobia can start to look less like benefactors and more like tenants demanding rent for a service no longer needed. The relationship can slide along the spectrum toward parasitism without either partner changing species, simply because the environment rewrote the accounting.

The fascinating part is that plants are not passive victims of this drift. Experiments by Toby Kiers and colleagues (2003) made the point sharply by showing that legumes can impose *host sanctions*: nodules containing ineffective rhizobia receive less oxygen and fewer resources, reducing the bacteria's success. This is not punishment in a moral sense; it is management. The plant is using physiology to audit a bargain. If the microbe is not delivering nitrogen, the plant cuts its losses. In the language of intimacy, sanctions are a way of keeping closeness from becoming exploitation.

A similar logic plays out underground, in one of the most widespread symbioses on Earth: mycorrhizae, the associations between plant roots and fungi. The fungus extends the plant's reach into the soil with fine filaments that can tap phosphorus and other nutrients beyond the immediate root zone. The plant pays in sugars produced by photosynthesis. For decades this was described as an almost pastoral mutualism: plants feed fungi, fungi feed plants, everyone prospers. Then careful experiments and isotopic tracers complicated the picture. The exchange is not always generous, and it is not always equal.

Kiers and colleagues (2011) showed that plants can reward fungal partners that deliver more nutrients by sending them more carbon. Fungi, in turn, can preferentially allocate nutrients to roots that supply more carbon. The result is not a single fixed bargain but a responsive market-like interaction, where the "price" of phosphorus in units of sugar can change with supply and demand. When soil phosphorus is scarce, fungal delivery is valuable, and plants are willing to pay more. When phosphorus is abundant, the plant's willingness to pay drops, and fungal strategies that extract carbon without providing much in return become more profitable. The relationship can tilt toward parasitism not because the partners are inherently treacherous, but because selection will favor whatever strategy yields net gain under current conditions.

This is where the chemistry of cooperation becomes visibly conditional. Carbon is not a vague "resource"; it is a particular set of molecules that must be transported across membranes. Nitrogen is not a favor; it is an energetically costly reduction reaction with vulnerabilities to oxygen. Phosphorus is not just "fertility"; it is phosphate ions that bind readily to soil particles and become hard to access. Symbiosis is not only about living together, but about the plumbing and the price tags that living together requires.

The intimacy spectrum also plays out at the scale of entire ecosystems. Coral reefs, for example, are built by animals that look like plants only because they are hosting algae. Reef-building corals contain symbiotic dinoflagellate algae (often discussed under the name *Symbiodiniaceae*) living inside their tissues. The algae photosynthesize and provide the coral with carbon-rich compounds; the coral provides shelter, carbon dioxide, and access to light. Under stable conditions this mutualism can be astonishingly productive, allowing reefs to flourish in nutrient-poor tropical waters that would otherwise support far less life. The vivid colors of corals come largely from the pigments of their algal partners.

Yet this is an intimacy that can fail dramatically. Under heat stress, intense light, pollution, or other disturbances, the delicate exchange can break down. The algae can become less efficient, generating damaging reactive oxygen species; the coral, attempting to protect itself, may expel the algae. The result is coral bleaching: the coral turns pale because the pigmented algae are gone. A bleached coral is not necessarily dead, but it is starving. The same close relationship that makes reefs possible also creates a point of vulnerability: when the partnership collapses, the coral loses a major energy source. Mutualism, in other words, is not a stable state of grace. It can be a high-performance arrangement that is sensitive to shocks.

Bleaching also illustrates a subtle point about the spectrum. If an algal symbiont under stress produces more damaging byproducts than useful sugars, then from the coral's perspective the symbiont has become costly. The coral's expulsion response can be seen as a kind of sanction, though the mechanism and stakes differ from legumes and rhizobia. What looks like a tragedy at the ecosystem level is, at the level of the individual coral, an attempt to manage an internal relationship whose costs have suddenly spiked.

The spectrum is not only between species. The same logic applies to our own bodies, where microbial partners can behave very differently depending on context. The human gut microbiome contains bacteria that help break down complex carbohydrates, produce vitamins, train immune responses, and exclude pathogens. Many of these microbes are best described as mutualists, or at least beneficial commensals. But context can shift quickly. A bacterium that is harmless in the gut can become dangerous if it crosses into the bloodstream or urinary tract. The microbe has not changed its identity; the host environment has changed the terms. The immune system, in turn, is not simply a war machine against microbes but a regulator of intimacy, tolerating many organisms in one compartment while treating the same organisms as threats in another.

These shifting categories suggest a more useful question than "Is this relationship good or bad?" The more revealing question is: *what controls the position on the spectrum?* Several recurring controls appear across systems.

One is *resource availability*. When a resource is scarce, a partner who provides it becomes valuable. When it is abundant, the same partner can become dispensable or even burdensome. Fertilizer can destabilize legume mutualisms for this reason. Nutrient enrichment in coastal waters can destabilize coral symbioses by promoting algal overgrowth and altering microbial communities.

Another control is *outside options*. If a host can choose among many potential partners, it can be picky and can reward the best. If partners are rare or transmission is strictly vertical (passed from parent to offspring), then choice is limited and other enforcement mechanisms become important. In mycorrhizae, plants often interact with multiple fungi and can, to some degree, shift carbon allocation. In highly obligate endosymbioses, choice is minimal; stability depends more on integrated control and shared fate.

A third control is *time*. Many relationships become more mutualistic when partners interact repeatedly and build mechanisms that reduce cheating. That idea has a formal life in evolutionary theory, but it also has a chemical life: repeated interactions are repeated exposures to signals. Over time, partners evolve receptors, transporters, and metabolic couplings that make the relationship more reliable. A bacterium that evolves to depend on a host-supplied nutrient is less likely to leave; a host that evolves to depend on a symbiont's vitamin production is less likely to expel it. Dependence tightens intimacy, but it also raises the cost of failure.

Historical perspective helps here, because the very words used to describe these relationships were forged in debate. The botanist Anton de Bary introduced "symbiosis" in the nineteenth century to describe organisms living together, not necessarily helpfully. Later, as microbiology and ecology matured, the terms mutualism and parasitism took on the sharper meanings used today. The shifting definitions reflect a shifting scientific mood: from seeing close associations as curiosities, to treating them as drivers of ecology and evolution, to recognizing that the same association can be both helpful and harmful depending on the circumstances. When Lynn Margulis argued for symbiogenesis, she was not only proposing a mechanism for the origin of complex cells; she was also pushing against a cultural preference for stories centered on competition. The spectrum of intimacy is, in part, a spectrum of scientific imagination: what kinds of explanations feel natural to offer.

A practical payoff of this way of thinking appears in medicine and conservation. Consider a bacterium like *Helicobacter pylori*, famous for causing ulcers and increasing stomach cancer risk, yet also increasingly discussed as a long-term inhabitant of humans that may modulate immune responses and influence disease risks in more complicated ways. Or consider fecal microbiota transplantation, which can treat recurrent *Clostridioides difficile* infections by restoring a microbial community that suppresses a pathogen. These interventions make little sense if microbes are only enemies. They make immediate sense if hosts are ecosystems whose health depends on managing a shifting set of internal relationships.

The spectrum also cautions against romanticizing mutualism. Mutualistic relationships can be stable, but they often require continuous regulation. They can also impose hidden costs. A plant supporting mycorrhizal fungi must divert carbon that could have gone to growth or reproduction. A coral hosting algae must tolerate a foreign cell within its own tissues, with the risk of oxidative stress. A human tolerating gut microbes must accept that immune defenses will be trained and sometimes mis-trained by microbial signals. Mutualism is not free. It is a strategy that pays when its costs are lower than the alternatives.

The deeper lesson is that the boundaries between parasitism, commensalism, and mutualism are not walls. They are more like colored zones on a weather map, shifting with pressure systems. The chemistry of cooperation is sensitive to gradients: nutrient gradients, oxygen gradients, temperature gradients, signaling gradients. Partners make decisions, in the only sense that biology permits, by responding to these gradients through gene regulation and metabolism. A host that "chooses" is often a host that differentially feeds; a symbiont that "cheats" is often a symbiont that diverts molecules inward rather than exporting them.

Seeing intimacy as a spectrum does not drain nature of drama; it sharpens it. The drama is not only who eats whom, but who tolerates whom, who invests in whom, and under what conditions a truce turns into a takeover. A reef is a monument to cooperation until a heatwave exposes how narrow the conditions for that cooperation can be. A field of legumes is a quiet bargain between plant and bacterium until fertilizer makes the bargain irrational. A forest is a web of plant—fungus trades until drought shifts the value of carbon and water, and the old exchange rates no longer hold.

In the end, the spectrum is not a philosophical compromise between good and bad relationships. It is a reminder that living systems are accountants of necessity, constantly recalculating. The sums are written not in numbers but in molecules: sugars moved through membranes, nitrogen reduced at an energetic cost, phosphate delivered from distant soil particles, reactive oxygen species held in check. Intimacy persists when the chemistry balances, and collapses when the balance tips too far, leaving partners to renegotiate their lives in the only currency that evolution recognizes.

1.3 The Ledger of Life

Cooperation in nature can look warm from a distance: a coral glowing with its algae, a root braided with fungal threads, a gut community quietly digesting what its host cannot. Close up, it looks less like a handshake and more like bookkeeping. Every molecule handed over is a molecule not used elsewhere. Every helpful enzyme costs energy to build and maintain. Every open doorway for a partner is also an open doorway for exploitation. The romance fades into a ledger: debits, credits, and the constant risk that someone will take the benefits without paying the costs.

The unsettling part is that the "someone" is not necessarily a different species. A cheater can be a mutant lineage within a cooperative population, a symbiont strain inside a host, or even a cell inside a multicellular body. The logic is the same at every scale. If a shared good is valuable and its production is costly, then individuals that avoid the cost while still receiving the good can outcompete the producers. Cooperation does not fail because nature is nasty; it fails because arithmetic is unforgiving.

It is useful to name the problem precisely. Many cooperative acts create what economists call a *public good*: something costly to produce, beneficial to many, and difficult to restrict to those who paid for it. In microbes, public goods can be digestive enzymes released into the environment, iron-scavenging molecules called siderophores, or the sticky polymers that hold a biofilm together. In a mutualism, the "good" might be fixed nitrogen leaked from a bacterial cell, phosphorus delivered by a fungal filament, or sugars exported by an alga. If access to the good cannot be controlled, freeloaders thrive.

That is the bleak baseline. The more interesting fact is that cooperation nevertheless persists across the living world, and often with remarkable stability. The reason is not that cheaters are rare. It is that cooperation evolves with countermeasures. Biology, lacking courts and contracts, relies on mechanisms that make payment hard to evade: partner choice, sanctions, screening, repeated interactions, and feedbacks that tie a partner's success to the success of the group it depends on. These mechanisms are the enforcement wing of the symbiosis engine, and they are often chemical at their core.

A historical anecdote shows how starkly this ledger can be written in theory. In 1964, W. D. Hamilton reframed social behavior with a simple inequality that made cooperation feel less mysterious. Helping a relative can evolve when the genetic relatedness between helper and recipient, multiplied by the benefit to the recipient, exceeds the cost to the helper. The appeal of Hamilton's approach was not that it made organisms selfish in a cynical way, but that it made them legible: a helper's genes can spread not only by producing its own offspring but by aiding copies of itself in relatives. Inclusive fitness, as Hamilton called it, is an accounting system where the currency is genetic replication. You can dislike the metaphor of "genes for altruism" and still appreciate the accounting move: cooperation is not outside natural selection; it is one of its possible outcomes when the numbers work.

Hamilton's insight does not explain all cooperation, especially between unrelated partners. For that, another elegant idea entered in 1971, when Robert Trivers described *reciprocal altruism*. If two individuals interact repeatedly, and if each can recognize or remember the other well enough to adjust behavior, then short-term generosity can pay long-term dividends. The problem, again, is cheaters: an individual can accept help and then vanish, or accept help and refuse to return it. Reciprocity survives only if there are ways to detect defection and respond by withholding future benefits. The social emotions that humans experience as gratitude, guilt, and anger can be seen, in this light, as psychological machinery for keeping ledgers. In other species, the machinery is often simpler, but the principle is recognizable: cooperation stabilizes when betrayal has consequences.

These theories are sometimes presented as grand abstractions, but they connect directly to the chemistry of symbiosis. Repeated interaction is, at the molecular level, repeated exposure to cues: surface proteins, secreted metabolites, signaling molecules. Recognition is receptor binding. Memory can be an altered pattern of gene expression, or a persistent change in a microbial community's composition. Consequences can be denial of nutrients, an oxidative burst, an immune response, a physical barrier. Nature can do enforcement without ever thinking about it.

The legume—rhizobium symbiosis, introduced earlier as an intimacy that can drift, is also a perfect case study in enforcement. Nitrogen fixation is the service, carbon is the payment, and the nodule is the marketplace. The potential cheat is obvious: a rhizobium strain can infect a root, take up residence in a nodule, and divert plant-supplied carbon to its own replication without fixing much nitrogen. Such a strain should, in principle, outcompete more cooperative strains, because it avoids the cost of running the nitrogen-fixing machinery. Left unchecked, the mutualism should decay.

Yet it persists, because hosts are not helpless. Kiers and colleagues (2003) demonstrated that legumes can reduce oxygen supply and other resources to nodules that are not fixing nitrogen effectively. Oxygen is a particularly powerful lever because nitrogenase, the key enzyme complex for nitrogen fixation, is damaged by oxygen, but the bacteria still need oxygen for respiration to generate ATP. The nodule is therefore engineered as a low-oxygen environment, with the plant controlling oxygen diffusion and using oxygen-binding proteins to buffer it. If the plant restricts oxygen further, it can choke off the energy budget of non-performing bacteria. In effect, the host keeps the books and can cut credit to a delinquent supplier.

Sanctions do not need to be perfect to matter. They only need to make cheating less profitable than cooperation. A cheater that replicates quickly but produces fewer successful descendants overall because its host starves the nodule is not a winning strategy. In this way, the plant can shape selection among rhizobia strains, nudging evolution toward reliability.

Partner choice is another stabilizer, and it looks even more like a market. If a host can choose among multiple potential partners, it can preferentially associate with those that offer better terms. Mycorrhizal symbioses provide a striking example because plants often interact simultaneously with multiple fungal networks. The plant has a currency (carbon) and can allocate it unevenly. The fungus has a currency (nutrients, especially phosphorus and nitrogen) and can allocate it unevenly. Kiers and colleagues (2011) showed that this bidirectional control resembles reciprocal rewards: better partners receive better payment. Outside options matter. A fungus competing with other fungi for plant carbon has an incentive to be generous enough to remain attractive; a plant with multiple fungi available has leverage to demand better service.

Biological market theory, developed by Ronald Noë and Peter Hammerstein in the 1990s, formalized this intuition: cooperation can be shaped by supply and demand when partner choice is possible. The "price" of a cooperative act depends on how many partners are available and how badly the commodity is needed. This is not to claim that plants calculate exchange rates, but to say something more concrete: evolution can tune sensory and transport systems so that resource allocation responds to partner performance in a way that mimics bargaining.

Seeing mutualisms as markets also clarifies why they are so sensitive to environmental change. Flood a soil with phosphorus fertilizer and the fungal commodity becomes less scarce; plants may pay less carbon, and fungi that extract carbon without delivering much nutrient can do better. Heat-stress a coral and its algae may become net producers of reactive oxygen species rather than net suppliers of sugars; the coral's best "deal" may become expulsion. Cooperation is not fragile because it is morally weak; it is fragile because the price of the commodity can crash.

The ledger metaphor can be made literal in a simplified balance sheet. Consider, very crudely, a plant engaging a rhizobial partner. The plant pays a carbon cost to build nodules and supply sugars; the rhizobium pays an ATP cost to fix nitrogen; both gain if the exchanged goods enable growth that outweighs these costs. Precise numbers vary by species and conditions, but the structure is robust: costs must be paid up front; benefits often arrive later; and the temptation to default is always present.

A simplified ledger: plant–rhizobium trade

Partner	Costs (paid)		Gains (received)	
Plant	Carbon to nodule	○○○	Fixed nitrogen	□□□
	Nodule construction	○○	Growth capacity	□□
Rhizobium	ATP spent fixing N	○○○○	Carbon from plant	○○○
	Risk of host sanctions	⚠	Protected niche	⌂

Ledger logic: cooperation persists when enforcement and choice make paying cheaper than cheating.

The point of such a cartoon ledger is not precision but clarity. Mutualism is not a cloud of goodwill; it is a trade where each side risks being left holding the bill. The stabilizers are the mechanisms that change the payoff structure so that reliable trade is the best available strategy.

Not all stabilizers look like sanctions or partner choice. Sometimes cooperation is stabilized by *partner-fidelity feedback*: a partner's success becomes automatically tied to the host's success because the partner cannot easily leave and find a better host. Many vertically transmitted symbionts fall into this category. If a bacterium is passed from mother to offspring, then its long-term prosperity depends on the host lineage flourishing. Cheating that harms host reproduction is self-defeating. This alignment can make cooperation stable even without active punishment, although it can come at a cost: such symbionts often lose genes for independent life and become hard to replace if conditions change.

Another route is *screening*, a term borrowed from economics and used in biology to describe situations where a host sets up conditions that only good partners can tolerate. Instead of inspecting each applicant, the host builds an environment that filters them. In the microbial world, where hosts often cannot "see" the internal state of a bacterium directly, screening can be more realistic than perfect partner choice. A host might, for example, create a chemically challenging niche that only symbionts with particular metabolic capacities can thrive in. The cost of entry becomes a test. A cheater that cannot pay the entry fee is excluded.

These ideas can sound abstract until one notices how often they rely on small molecules. Screening can be oxidative stress, acidity, or antimicrobial peptides. Sanctions can be withholding oxygen, carbon, or access to a protected compartment. Rewards can be preferential sugar export, altered membrane permeability, or growth factors. Even in animal societies, the currency of cooperation often becomes chemical: food sharing is the transfer of calories; grooming is parasite removal plus hormonal effects; territorial defense is energy expenditure and injury risk. Underneath, the ledger is always metabolic.

The ledger view also helps with a recurring confusion about cooperation: the assumption that it must be explained by some special force opposed to natural selection. In practice, the explanation usually lies in the details of who interacts with whom and who can control access to resources. When interactions are structured—by kinship, by spatial clustering, by repeated encounters, by the architecture of a host body—cooperation can be locally stable even if it seems globally fragile. Microbes in a biofilm, for example, often interact mostly with close neighbors, meaning that a producer's benefits are more likely to help relatives or at least nearby cooperators. A cheat might gain locally, but it can also destabilize the microenvironment it depends on, leading to collapse. Stability can emerge from geometry.

This is also why debates in evolutionary biology about how best to frame cooperation can become intense. Different mathematical languages—kin selection, multilevel selection, game theory on networks—often describe overlapping realities from different angles. Nowak's synthesis of "rules" for the evolution of cooperation (2006) popularized a broad map of stabilizing mechanisms, while subsequent debates, including high-profile disputes over the foundations of inclusive fitness, reminded everyone that the bookkeeping can be done with different ledgers. The living world does not care which notation scientists prefer; it only cares that the costs get paid.

There is a final twist that brings the story back to chemistry rather than mathematics. Cheating is not always a conscious strategy; in microbes it is often a metabolic mutation. A strain that fails to produce a costly public good may simply have lost a gene, and the loss is favored because it saves energy. Enforcement, similarly, is often not a moral response but a physiological reflex. A legume nodule starved of oxygen is not staging a tribunal. A plant root allocating less carbon to a poorly performing fungus is not nursing resentment. These systems work because molecules are harder to lie with than words. If nitrogen does not arrive, the plant's nitrogen sensors register deficiency. If phosphorus delivery is low, growth slows, and carbon allocation shifts. The "audit" is embedded in the metabolism itself.

Cooperation, then, is a way of arranging chemical constraints so that honesty is cheaper than deception. The most successful mutualisms do not require saintly partners; they require incentives aligned tightly enough that even self-interested lineages, shaped only by differential reproduction, behave as if they were trustworthy traders. The ledger does not eliminate intimacy. It explains how intimacy can survive contact with reality: not by ignoring conflict, but by pricing it in.

Chapter 2

Molecules of Meaning

To the human eye, a coral reef or a forest canopy appears as a collection of discrete physical entities, yet the space between them is thick with invisible information. By tuning into the frequency of chemical ecology-the study of nature's oldest language-we discover that cooperation is not merely an instinct but a negotiated dialogue mediated by complex molecules. This exploration reveals how organisms shout, whisper, and listen in a chaotic environment, and why the high metabolic cost of these chemical words ensures they are rarely spoken in vain.

2.1 Whispers in the Water and Air

A walk through a pine forest feels quiet because human ears are the wrong instrument. The air is thick with messages that never become sound: resinous monoterpenes leaking from needles, green-leaf volatiles released by a twig snapped underfoot, faint aromatic cues that drift ahead of an approaching insect or linger after an animal has passed. On a coral reef the situation is even more extreme. Seawater is not an empty medium but a slow, swirling archive of molecules: amino acids, sugars, sulfur compounds, and countless metabolites shed, exuded, or actively secreted. Life moves through these spaces as if through a library whose books dissolve and re-form every second.

Chemists and biologists now take it almost for granted that organisms communicate with chemicals. Yet this idea had to be fought for. For much of modern biology, the senses worth dignifying were the ones we share most obviously with other animals: vision and hearing. Smell was treated as a supporting actor, taste as a footnote, and anything happening below the threshold of our own perception as biological trivia. The world was assumed to be, if not silent, at least chemically unremarkable.

The field that changed that sensibility is often called *chemical ecology*: the study of how natural chemicals mediate interactions between organisms. It is not simply about toxins and nutrition. It is about information. A molecule can be food or poison, but it can also be a sentence. In this view, an organism does not merely occupy an environment; it writes into it, reads from it, and learns to distinguish meaningful phrases from background noise.

Two scientists did as much as anyone to make this view irresistible: Thomas Eisner, a biologist with an experimentalist's playfulness, and Jerrold Meinwald, a chemist with an artist's feel for structure. Their partnership, forged at Cornell University in the middle decades of the twentieth century, gave chemical ecology a distinctive style: the boldness to ask what an organism is *doing* with a chemical, paired with the analytical care to identify exactly what that chemical is. They made it normal to picture a naturalist kneeling in the field with forceps and vials, and then imagine the same naturalist walking into a laboratory where glass columns, spectrometers, and painstaking structure determination would translate an animal's behavior into molecular terms.

Their work is full of memorable episodes, but one captures the spirit especially well. Eisner was fascinated by insects that defend themselves with chemistry, not as abstract curiosities but as living demonstrations that molecules can shape destiny. In many insects, the defensive secretion is not sprayed casually; it is released at the right moment, in the right direction, often with startling precision. Eisner would provoke these defenses, collect minute droplets on glass, and then ask Meinwald to do the hard work of identification. What emerged from collaborations like this was not merely a catalog of exotic compounds. It was a new way of thinking about an interaction. A predator's hesitation, a prey's survival, a community's structure: each could hinge on a chemical signature.

The most famous defensive spectacle is the bombardier beetle, which ejects a hot, noxious spray generated by mixing reagents in a specialized chamber. The drama is cinematic: a small beetle producing a burst of irritating chemicals with audible pops. What made this kind of system so valuable scientifically was not only its theatricality but its logic. The beetle's defense is an *act*, timed and aimed, and that makes it hard to dismiss as mere metabolic waste. It invites the question of meaning: what does this secretion *say* to a predator? Not "I exist," but "I am a costly mistake."

To talk about chemical meaning without drifting into metaphor, it helps to name categories. The broad umbrella term *semiochemical* (from the Greek *sēmeion*, a sign) refers to chemicals that carry information between organisms. Within that umbrella, the first major split is between signals used within a species and cues that cross species boundaries.

A *pheromone* is a semiochemical used for communication within a species. It is often portrayed as a simple "sex scent," but pheromones do much more: they can mark trails, signal alarm, coordinate development, or mediate aggression. They are typically "private" in the sense that they have been shaped by selection to work reliably between members of the same species, often through specialized receptors tuned to the right molecular shapes.

When the chemical crosses species lines, the terminology shifts. Many chemical ecologists use *allelochemical* for a semiochemical that mediates interactions between different species. Allelochemicals can be subdivided by who benefits. An *allomone* benefits the emitter (a repellent that discourages predation, for example). A *kairomone* benefits the receiver (a predator that detects a prey's odor, or a parasite that homes in on a host). A *synomone* benefits both parties (a plant volatile that attracts an insect that will pollinate it; a cue that draws a predator to an herbivore that harms the plant).

These categories can sound like taxonomic fussiness until one remembers what is at stake. In nature, information is rarely free. If a scent helps a mate find you, it can also help an enemy find you. If a chemical deters a predator, it might also advertise you to a specialist that has evolved to tolerate it. The "meaning" of a molecule is not fixed. Meaning is negotiated in the tug-of-war between emitters and receivers, and it can be revised by evolution whenever one side gains an advantage.

The air makes this negotiation fast and expansive. Volatile organic compounds travel quickly, especially when warm air rises and wind shears plumes into filaments. A plant does not need to touch an herbivore to announce that it is under attack; damage can release a bouquet of volatiles that drift outward and upward. Even a human can sometimes sense the sharp, green odor of crushed leaves: that smell is not a botanical perfume meant for us, but part of a chemical language that long predates our species.

One of the most elegant examples of airborne chemical messaging links three players: a plant, the insect that eats it, and the insect that eats the eater. When caterpillars chew leaves, some plants respond by releasing a blend of volatiles that attracts parasitic wasps. The wasps use the plant's odor plume as a locator beacon, find the caterpillar, and lay eggs that will consume it from within. From the plant's perspective, the wasp is an ally; from the caterpillar's perspective, the plant has just betrayed it. The same chemical plume can be understood as a synomone (plant and wasp benefit) and, in a darker register, as a weaponized advertisement.

Water changes the geometry of whispers. Molecules diffuse more slowly in water than in air, and currents can carry solutes in coherent trails that persist. Many aquatic organisms exploit this. A predator can follow the chemical track left by prey; prey can detect alarm cues released by injured individuals; larvae can recognize a suitable habitat by its dissolved chemical signature. The ocean is sometimes described as dilute, but for an organism equipped with sensitive receptors, dilution is not silence. It is a different kind of sentence: one written in gradients.

The reef is a particularly rich place to think about chemical meaning because so much depends on recognizing the right partners. Corals, for example, are not solitary animals in a simple sense. Many reef-building corals rely on symbiotic algae (often called zooxanthellae) living within their tissues, providing photosynthetic products that help fuel the coral's growth. To form and maintain such associations, organisms must distinguish friend from foe, tenant from pathogen, useful strain from useless one. That recognition begins, often, as chemistry: surface glycans that act like identity badges, small metabolites that bias immune responses, dissolved cues that influence settlement.

A similar logic plays out on land, in soils where roots and microbes meet. The ground is not merely a substrate. It is a molecular commons filled with exudates from roots, metabolic byproducts from microbes, fungal secretions, and decomposing organic matter. In this soup, a plant has to recruit helpful partners without inviting opportunists that will chew its tissues or drain its resources. The fact that such negotiations can occur at all depends on exquisite chemical specificity, but even before specificity there is ubiquity: the mere fact that *everything* leaks molecules, and that evolution can turn leakage into language.

Eisner and Meinwald helped biologists see that ubiquity without becoming numb to it. Part of their genius was methodological: they treated nature's small dramas as solvable mysteries. An insect makes an ant recoil. A beetle survives an attack from a bird. A butterfly is avoided by predators despite being conspicuous. Instead of stopping at "interesting behavior," they asked: what compound causes it? how is it produced? where is it stored? what does it cost? how does it change the ecology of the organism? In doing so, they implicitly argued that the meaning of a molecule is not a poetic overlay; it is a measurable property, grounded in receptors, doses, and consequences.

That attention to dose is crucial. Chemical communication often works at concentrations so small they seem unbelievable. An alarm pheromone may be effective in trace amounts. A marine larva may respond to nanomolar levels of a dissolved cue. This does not mean that organisms possess supernatural sensitivity; it means that their sensory machinery has been tuned by selection to detect particular molecules against noisy backgrounds. For humans, a whiff of smoke can dominate a room because our brains have been trained (culturally and biologically) to treat it as urgent. For many insects, a particular aldehyde or ester can dominate their perceptual world in the same way. The "loudness" of a chemical signal is not its mass but its relevance to a receptor and a nervous system.

Noise, however, is always present. Forest air is full of terpenes and microbial volatiles; seawater is full of metabolites from plankton and bacteria; soils are chemical tapestries woven by countless organisms.

If a message is to be meaningful, it must be distinguishable. Sometimes this is achieved with rarity: a molecule that few other organisms make. Sometimes with ratios: a distinctive blend whose components are common but whose proportions are informative. Sometimes with timing: a burst released only during an attack, or only at night, or only when an organism is reproductively ready. Nature often writes in mixtures, not single compounds, because mixtures are harder to counterfeit and easier to encode with context.

There is also a social dimension: chemical information can be exploited. A pheromone that calls a mate can be intercepted by a predator. A defensive secretion that repels a generalist predator can attract a specialist adapted to that chemical. A plant volatile that signals herbivory can alert neighboring plants, but it can also alert more herbivores to a palatable meal. Chemical landscapes are "public" in a way that sound and light are public, but with an additional twist: the messages can stick to surfaces, dissolve into fluids, and be sampled long after the sender has moved on. A footprint is not just a physical indentation; it is a residue of skin lipids and microbial metabolites that can be read as presence, direction, and even identity.

Human beings participate in these networks whether we intend to or not. Agriculture, for example, is an enormous experiment in chemical ecology. When we plant large monocultures, we create vast, coherent odor landscapes that herbivores can locate easily. When we breed plants for certain traits, we may inadvertently alter their volatile blends, changing how pests and beneficial insects respond. The idea of using pheromones and other semiochemicals for pest management arises directly from the recognition that behavior can be steered by molecules rather than bludgeoned by broad-spectrum poisons. This is not merely a "green" alternative; it is an intellectual shift. Instead of trying to dominate an ecosystem with force, one can try to speak in its native chemical dialect.

Chemical ecology also reshapes how we think about individuality. A species is not just a genetic lineage but a chemical presence. Many organisms carry a "cloud" of associated microbes that contribute to their odors and secretions. A smell can be an organism's own product, a microbial product, or a hybrid. That complicates the idea of a single sender and a single message. Yet it also enriches it. A signal may encode not only "I am here" but "I am healthy," "I have recently fed," "I belong to this colony," or "I carry these symbionts." Meaning becomes layered, like harmonics in music, except the harmonics are molecular.

Eisner and Meinwald's legacy is not only the compounds they identified but the stance they modeled: a kind of respectful curiosity toward the chemical lives of other organisms. They treated insects not as simple automatons but as chemists and listeners, producing and interpreting molecules with consequences as real as any spoken word. The world they revealed is one in which every boundary between organisms is also a contact zone between chemistries, and every habitat is a circulating medium of signs.

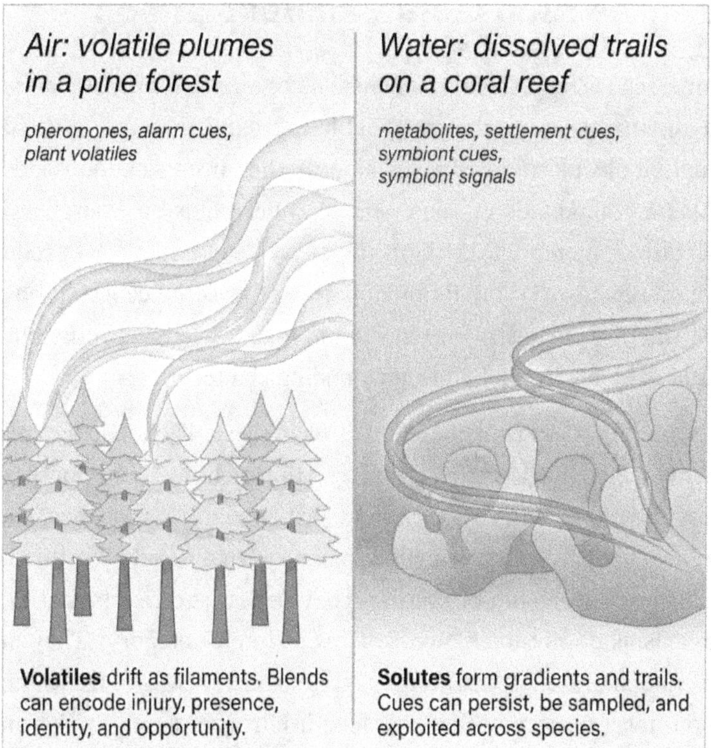

To say the biological world is never silent is not to romanticize it. It is to recognize that organisms, over evolutionary time, have turned the unavoidable fact of chemistry into a medium for choice. They leak molecules because they must; they learn to read those molecules because doing so changes who eats, who mates, who helps, who harms, and who survives. In air and in water, whispers are not a decorative feature of nature. They are one of its operating systems.

2.2 The Architecture of Recognition

A message is only useful if it can be recognized as *the right* message. In a forest or a reef, the chemical background is so crowded that a single molecule rarely arrives alone. The air carries thousands of volatile compounds at once, and seawater holds an ever-changing mixture of metabolites, fragments of cells, and microbial products. For an organism trying to find a partner, this is the central problem: not whether chemistry can travel, but how to read a specific phrase without being fooled by the surrounding chatter.

Recognition begins at the surface of bodies. If chemical ecology paints the environment as an information-rich medium, physiology provides the grammar: receptors that bind certain shapes, membranes that gate entry, signaling pathways that decide whether contact means welcome or alarm. The deepest surprise is that many organisms do not treat "non-self" as a simple category. They must tolerate and even cultivate the right outsiders. A root has to make room for a bacterium that will feed it nitrogen; a coral polyp must accept algae into its own cells; an animal gut must allow a dense microbial community to settle without constant inflammation. Partnership is not the absence of immunity, but a more demanding kind of immune intelligence.

The popular image of recognition is the *lock-and-key* metaphor: a molecule (the key) fits into a receptor (the lock), triggering a response. It is a useful starting point, but it hides the real artistry. In living systems, locks are rarely rigid and keys are rarely perfect. Many receptors are flexible enough to bind a family of related molecules, yet specific enough to avoid a disastrous false positive. More importantly, binding is only the first step. A cell must decide what the binding *means*. The same general class of microbial molecule can be interpreted as a warning in one context and as a passport in another.

Plants offer a clean way into this problem because they do not have circulating immune cells that roam the body; each plant cell must defend itself, while the whole organism coordinates these local decisions across tissues. In soils, this becomes a high-stakes negotiation. The plant sits still while a multitude of microbes arrive, including pathogens, decomposers, harmless hitchhikers, and potential mutualists. If the plant rejects everyone, it may survive but stay starved. If it accepts too easily, it becomes lunch.

Among the best-studied mutualisms in biology is the partnership between legumes (peas, beans, clover, and their relatives) and nitrogen-fixing bacteria called rhizobia. Nitrogen is abundant in the atmosphere but hard for most organisms to use directly; it must be converted into ammonia or related compounds. Rhizobia can do this conversion, but the chemistry is costly and oxygen-sensitive. The solution, evolved many times within legumes, is to build special root organs called nodules that house the bacteria in a carefully controlled environment. Inside the nodule, the bacteria fix nitrogen; the plant supplies carbon and maintains conditions that protect the key enzyme systems. This is cooperation, but it is also architecture: a living apartment complex constructed in response to chemical introductions.

The first introductions happen long before any bacterium crosses a plant cell wall. Legume roots leak an assortment of compounds into the soil, including flavonoids—small, ring-shaped molecules that plants make for many reasons, from pigment to defense. In this context, some flavonoids serve as a selective broadcast. Certain rhizobia respond by turning on genes needed for symbiosis, including the machinery to produce a distinctive set of signal molecules called Nod factors.

Nod factors, also known as lipochitooligosaccharides, sound like the sort of thing only a biochemist could love. Their basic structure resembles short chains of N-acetylglucosamine (a sugar building block also found in chitin), decorated with a fatty acid tail and additional chemical modifications that vary among rhizobial strains. Those decorations matter. They are not merely ornamental. They help determine which legume species will respond, and how strongly.

The plant, in turn, carries receptors on its root cells that can detect Nod factors at extraordinarily low concentrations. These receptors are proteins embedded in the cell membrane, with external domains that bind the Nod factor and internal regions that trigger signaling cascades. Many of these receptors belong to a class known as LysM receptor-like kinases, named for a domain (LysM) that can bind certain sugar-containing molecules. The important point is not the acronym but the principle: the plant uses a receptor architecture that can *read* a subtle chemical signature amid the general molecular noise of soil.

That is already impressive. What makes the story more emotionally resonant, if one allows oneself to feel for a root, is the balancing act. The same soil contains fungi and bacteria that carry chitin-like or peptidoglycan-like molecules—molecular patterns that plants often treat as danger signals. A plant that reacts too strongly to anything resembling chitin will miss its chance at symbiosis. A plant that reacts too weakly will invite pathogens. Recognition therefore is not simply about detecting a molecule; it is about steering the immune system toward a particular interpretation.

The early stages of rhizobial recognition are full of physical drama at microscopic scale. When the right rhizobia are present and Nod factors are perceived, root hairs can curl, trapping bacteria in a loop of plant tissue. The plant then creates an infection thread—a narrow tube-like structure—through which bacteria can travel inward, guided into developing nodules. This is a controlled invasion, almost a paradox: the plant actively constructs a route for microbes to enter. It does so because the chemical exchange has crossed a threshold from suspicion to trust.

There is a historical beauty to how these mechanisms were uncovered. The basic phenomenon of nodulation was known long before the molecular details; farmers had long observed that legumes improve soil. The puzzle for scientists was how specificity arose. Why do certain rhizobial strains nodulate certain legumes but not others? The answer, gradually assembled over decades of genetics, chemistry, and physiology, was that specificity is encoded in both the *signal* and the *sensor*. Change the decorations on the Nod factor, and the plant may no longer respond. Change the receptor, and the same Nod factor may become invisible or even misread. Experiments transferring or modifying receptor genes showed that host range can sometimes be extended, a striking demonstration that compatibility can hinge on molecular fit rather than on any vague notion of "species harmony."

The rhizobium example can tempt one into imagining recognition as a tidy code: flavonoid out, Nod factor back, door opens. Real soils are messier. Roots exude compounds continuously, microbial communities shift with moisture and season, and plants face trade-offs between defense and growth. Yet the system works often enough to have reshaped ecosystems. A biochemical handshake, reinforced by receptor architecture and cellular decision-making, can redirect global nitrogen cycles.

In the sea, the problems of recognition are both similar and more intimate. A coral polyp is an animal that lives surrounded by microbes and bathed in dissolved organic matter. Many corals form partnerships with dinoflagellate algae, which can live inside the coral's cells. This is not merely cohabitation; it is endosymbiosis, an arrangement that requires the host cell to accept a foreign organism into a compartment without destroying it.

At first glance, this seems to contradict the logic of immunity. Animal immune systems are often described as self/non-self discriminators, primed to attack invaders. In practice, animal immunity is better understood as a manager of relationships. It must decide whether a microbial presence is damaging, harmless, or beneficial. The same molecular patterns that help detect pathogens can be used to manage symbionts, with context and downstream signaling making the difference.

Coral recognition involves a range of cues, but one particularly evocative mechanism draws on a familiar biochemical language: sugars and lectins. Glycans—complex sugar patterns on cell surfaces—are among the most information-dense molecules in biology. They can vary in branching, linkage, and decoration, creating an enormous vocabulary of shapes and textures. Lectins are proteins that bind specific glycan motifs, essentially serving as readers of this sugar code.

Laboratory studies have shown that altering algal surface glycans can disrupt successful colonization, suggesting that these sugars act as identity tokens. The host, for its part, produces lectin-like molecules that can bind compatible algal partners and influence uptake. More recent gene-level work strengthened this picture by linking specific host lectins to the processes of binding, phagocytosis-like engulfment, and immune modulation during early establishment. The details differ among coral species and algal strains, but the underlying idea is strikingly general: recognition can be a matter of molecular texture, read by proteins evolved to sense it.

The moment of uptake is especially poignant. A coral cell engulfs the alga, enclosing it in a membrane-bound compartment. In many other contexts, engulfment would be the prelude to digestion. Here it becomes the prelude to collaboration. To accomplish that reversal, the host must adjust the chemistry of the compartment and its immune signaling, preventing the usual destructive pathways while still keeping control. The symbiont, meanwhile, must avoid triggering host defenses and must function in a way that benefits the host rather than behaving as a parasite. All of this depends on an initial act of identification and a continuing process of monitoring.

This is where the lock-and-key metaphor becomes too thin. Locks do not typically have moods. Cells do. Binding events are interpreted through networks of signaling proteins, gene regulation, metabolic state, and stress responses. A host under thermal stress may interpret the same symbiont differently. A symbiont under nutrient limitation may shift its metabolism and become less generous. Recognition is therefore not a one-time passport check; it is closer to a continuous conversation in which both partners reassess.

The idea that innate immune receptors can promote symbiosis as well as inflammation helps make sense of this. Many animals possess pattern-recognition receptors (PRRs), proteins that detect common microbial features. In textbook narratives, PRRs trigger defense. In living tissues, PRR signaling can also support tolerance and development. In the gut, for example, microbial signals help shape the immune system and maintain barrier function. In symbioses, similar logic applies: microbial ligands can be interpreted not as threats to eradicate but as cues to manage, especially when the host receives reliable benefits.

A useful way to think about recognition is as a series of filters, not a single gate. The first filter is physical: can the potential partner reach the host tissue and remain there long enough to be sampled? The second filter is chemical: do the molecules presented resemble those associated with compatible partners? The third filter is physiological: does the host's internal state allow partnership now? The fourth filter is behavioral or developmental: will the host commit resources—build a nodule, engulf a symbiont, alter its immune tone—based on the information received? At each step, mistakes are possible, and the costs of mistakes shape how strict or permissive the filters become.

Modern metabolomics has begun to reveal how rich these chemical interactions are. Metabolomics is, in essence, large-scale measurement of small molecules in a system: a chemical census of tissues, exudates, or environmental samples. With sensitive instruments, researchers can compare the chemical profiles of hosts exposed to different microbes, track shifts over time, and identify candidate molecules involved in attraction, suppression, or activation. The power of this approach is that it respects the messy reality of biology. It does not assume there is only one signal. It allows for blends, context dependence, and unexpected intermediates.

At the same time, metabolomics can tempt one into a new kind of oversimplification: the belief that if a molecule changes, it must be a message. Cells produce and consume metabolites for many reasons. Some chemical cues are intentional signals shaped by selection; others are byproducts that happen to be useful to an eavesdropper. A root exudate might be a deliberate recruitment tool, or it might be a metabolic leak that microbes exploit. The boundary between signal and cue is often blurred, and organisms can evolve to take advantage of whatever is available. Recognition machinery does not necessarily care whether a molecule was produced "for communication"; it cares whether the molecule reliably predicts something important.

That reliability brings us back to architecture. Receptors are not passive locks. They are investments. A protein that binds a narrow range of molecules with high sensitivity is expensive to evolve and maintain, and it can become a liability if it is easily fooled. The shape of a binding pocket, the flexibility of a receptor domain, the downstream signaling pathways that interpret binding: these are all parts of a recognition device tuned by selection pressures in a chemical world full of deception, coincidence, and opportunity.

The emotional core of these systems is that they allow intimacy without surrender. A legume invites bacteria into specialized organs but can restrict resources to nodules that do not perform well. A coral may accept symbionts but can regulate their abundance and, under stress, expel them—a drastic act known to us because of coral bleaching. These are not harmonious mergers. They are controlled relationships maintained by constant sensing. Recognition is the beginning of that control, but also its continuing foundation: a way of knowing who is there, what they are doing, and whether the partnership is still worth the cost.

In a human life, recognition is often portrayed as instantaneous—love at first sight, or a snap judgment of character. In biology, recognition is usually slower, layered, and conditional. Molecules are sampled repeatedly, contexts are compared, internal states are integrated. The result is not certainty but workable confidence, sufficient to commit resources without being naïve. A root hair curling around a bacterium and a coral cell accepting an alga are both acts of discernment performed without eyes or ears. They rely on surfaces that feel chemistry the way fingertips feel texture, turning a turbulent molecular world into a readable map of possible allies.

2.3 The Currency of Trust

Cooperation has a publicity problem. The very acts that make partners valuable to one another often leave traces that can be exploited by bystanders. A root that opens its tissues to welcome a bacterium risks letting in a pathogen. A coral cell that relaxes its defenses to accommodate an alga risks becoming a hotel for something less helpful. A bacterium that secretes an enzyme to break down food for the community creates a resource that any nearby cell can steal. Chemical signals sit right in the middle of this dilemma, because they are both invitation and evidence: they announce intent, but they also reveal opportunity.

If signals were free, they would be worthless. Any organism could claim to be a cooperative partner. In human terms, it would be like a marketplace where every product label could be faked at zero cost, and the seller faced no penalty for fraud. Such markets collapse unless there is some mechanism that links what is said to what is true. In nature, that mechanism is often chemistry itself: the cost of making a molecule, the physiological constraints that couple signal production to genuine cooperative capacity, and the ability of receivers to test, punish, or abandon partners who do not deliver.

This is the territory of evolutionary economics. Not economics in the sense of money, but in the sense of budgets and incentives. Every cell and organism has limited resources and competing demands. A signal can be thought of as a kind of spending. It consumes energy, raw materials, time, or risk. The benefit is access: to a mate, to a host, to protection, to a nutrient stream. The key question becomes: under what conditions does honest signaling remain stable when cheating is possible?

A classic answer comes from Amotz Zahavi's handicap principle (1975), proposed to explain why extravagant animal displays can be reliable. The idea is disarmingly simple: if a signal is costly enough that only high-quality individuals can afford it, then the signal can remain honest because impostors would pay too much. The peacock's tail works, in this logic, because it is an expensive inconvenience. Chemical signals can function as handicaps too, not because *molecules are always difficult to synthesize*, but because *some* of the most informative molecules are metabolically demanding, require rare precursors, or impose collateral costs such as increased visibility to enemies.

Yet cost alone is not a magic wand. Even in human markets, an expensive advertisement does not guarantee a good product; it only guarantees that someone could pay for an advertisement. In biology, a high-cost signal can still be faked if the benefit of deception exceeds the cost, or if the cost is not actually linked to the cooperative act being advertised. So natural systems often rely on a more layered solution. The signal may be costly, but it may also be *constrained*: tightly coupled to underlying physiology so that faking is difficult. And the receiver is rarely passive. Partner choice and sanctions are the enforcement mechanisms that turn chemical talk into something like a contract.

Microbes provide an especially crisp laboratory for these ideas, because microbial cooperation often takes the form of secreted public goods. A bacterium might release enzymes that digest complex nutrients, siderophores that bind iron, polymers that build the matrix of a biofilm, or signaling molecules that coordinate collective behavior. These secretions are costly, and they can be exploited. A "cheater" strain can reduce or eliminate production while still benefiting from what others release.

Many bacteria use quorum sensing to manage this dilemma. Quorum sensing is a chemical communication system in which cells produce small signal molecules that accumulate as the local population density increases. When the signal concentration crosses a threshold, cells change gene expression, often turning on cooperative traits that are only worthwhile when many neighbors are present. The intuitive appeal is obvious: do not pay the cost of cooperation unless enough partners are nearby to make the investment pay off.

The social vulnerability is equally obvious: if a cooperative trait is turned on in response to a shared signal, then a cell that listens but does not contribute can free-ride. Experiments have demonstrated this logic in real microbial communities, including biofilms where cells live in dense, structured neighborhoods. Quorum-sensing-defective mutants can act as cheats, exploiting the shared environment without paying into it, and their presence can measurably degrade the collective phenotype. Biofilms can become thinner or less productive, and their resilience can drop in ways that matter outside the lab, including altered susceptibility to antibiotics. The chemistry that coordinates cooperation also creates a point of attack.

One might expect such systems to be evolutionarily fragile, yet quorum sensing is widespread. Part of the explanation is that microbial communities are not well-mixed soup. In a structured environment, cooperators can cluster with kin, meaning that the benefits of secretion return disproportionately to relatives who share the genes for secretion. Another part is enforcement through ecology: cheats may do well in the short term but poorly in the long term if the collective function collapses. A cheat can flourish in a biofilm built by cooperators, but if too many cheats accumulate, there may be no biofilm worth exploiting.

There is also a subtler mechanism that brings us closer to the idea of chemical honesty: sometimes the signal is not an arbitrary announcement but an unavoidable cue of investment. Modeling work has emphasized that if a signaling molecule is tightly linked to cooperation—for example, if it is a byproduct of the metabolic pathways that produce the public good—then lying becomes hard. A cell cannot cheaply produce the signal without also running the machinery that makes cooperation possible, or at least incurring similar costs. In that case, the signal is more like smoke from a fire than a deliberate flag: if you see it, you can infer something real is happening.

This cue-versus-signal distinction matters because it reframes what "honesty" means. Honesty is not always maintained by punishment; sometimes it is maintained by constraint. A host does not need to threaten a microbe with sanctions if the microbe cannot easily mimic the biochemical signature of cooperation without actually cooperating.

Still, sanctions exist, and they can be remarkably direct. Consider again the legume–rhizobium partnership. The plant supplies carbon to bacteria in nodules, and in return it expects fixed nitrogen. But not all rhizobia are equally generous, and some can behave as poor partners. Plants are not helpless in this arrangement. They can reduce oxygen supply or otherwise limit resources to nodules that do not provide enough nitrogen, effectively imposing a penalty. The plant cannot read bacterial intent directly, but it can measure outcomes and adjust payment accordingly. This is a biological version of performance-based contracting: compensation is linked to delivery.

Partner choice operates on similar logic but earlier in time. If a host can preferentially admit certain partners—those with the right chemical signatures, those that trigger the right developmental pathways, those that match local ecological conditions—then cheats and low-quality partners are filtered out before they gain access to resources. Recognition mechanisms set the stage, but the currency of trust is paid in what happens after recognition: continued access, continued nourishment, continued shelter.

The chemistry of signals is central to this currency because complex signals tend to be harder to forge. In the same way that it is easier to counterfeit a plain stamp than a sophisticated watermark, it is generally easier to mimic a simple molecule than a precise blend with multiple components, stereochemical specificity, and tightly controlled timing. Complexity can raise the barrier to entry for cheaters. It does not guarantee honesty, but it changes the economics.

There is, however, a trade-off. Complexity is often expensive not only to produce but also to interpret. Receivers need receptors and processing pathways to distinguish fine chemical differences. This is one reason natural systems often reuse familiar molecular building blocks—sugars, lipids, amino-acid derivatives—and then add small modifications to encode specificity. A modest chemical tweak can create a new "dialect" without reinventing the whole sensory apparatus. In rhizobia, for example, variations in Nod factor decorations can alter host range. The plant's receptors must be sensitive to these variations, but they do not have to start from scratch.

Costs can also be less obvious than ATP expenditure. A volatile signal can draw in allies, but it can also betray location. An organism that produces a conspicuous chemical trail may gain a partner more quickly but may also increase predation risk. Those risks, too, are part of the budget. When chemists speak of "metabolic cost," it is easy to picture a ledger of carbon atoms and energy units, but the ecological costs can be equally decisive. The world listens.

Plant chemistry provides vivid examples of costly messaging that doubles as defense. Glucosinolates, the sulfur-containing compounds famous in mustards and cabbages, are not mere toxins. They are part of a broader chemical strategy that includes signaling, deterrence, and the management of microbial communities around roots and leaves. Synthesizing such compounds demands nitrogen and sulfur, and their production can impose measurable energetic burdens, including increased requirements for photosynthetic input. Even without knowing the details, one can appreciate the logic: if a plant invests heavily in such chemistry, the investment itself can make certain signals more reliable. A weak plant cannot casually pretend to be chemically fortified.

Yet biology is rarely satisfied with one enforcement mechanism. Chemical costs, constraints, partner choice, and sanctions often work together. A useful way to picture this is as an economy with multiple layers of security. A bank does not rely only on a lock; it uses locks, alarms, cameras, audits, and legal penalties. In symbiosis, locks and cameras are receptor specificity and chemical complexity; audits are physiological monitoring; penalties are sanctions such as reduced resources or expulsion.

Microbial communities add another twist: heterogeneity. Real populations are not uniform cooperators or uniform cheats. They often contain mixtures, and that mixture can persist. Behavioral variation in quorum-sensing responses, for example, can stabilize coexistence between cooperators and defectors. At first glance, this seems like failure—why tolerate cheats at all? But a certain level of diversity can be a form of insurance. If conditions change, the population has multiple strategies ready. And if cooperation is beneficial but only under certain circumstances, then having some cells that refrain can prevent wasteful overinvestment.

This resembles a familiar human pattern: not everyone contributes equally to a public good at all times, yet the system can function if enough do, and if non-contributors do not become overwhelming. Biology often aims not for moral purity but for dynamical stability.

The following figure distills these ideas into a conceptual curve. It is not a law of nature but a way of seeing the incentives. Cheap, simple signals are easy to produce and easy to mimic, which can raise the likelihood of cheating. As signal complexity and cost increase, cheating tends to become less profitable, either because faking is too expensive or because the signal is more tightly coupled to genuine cooperative machinery. The curve does not hit zero: deception is a perpetual possibility, and enforcement is never perfect.

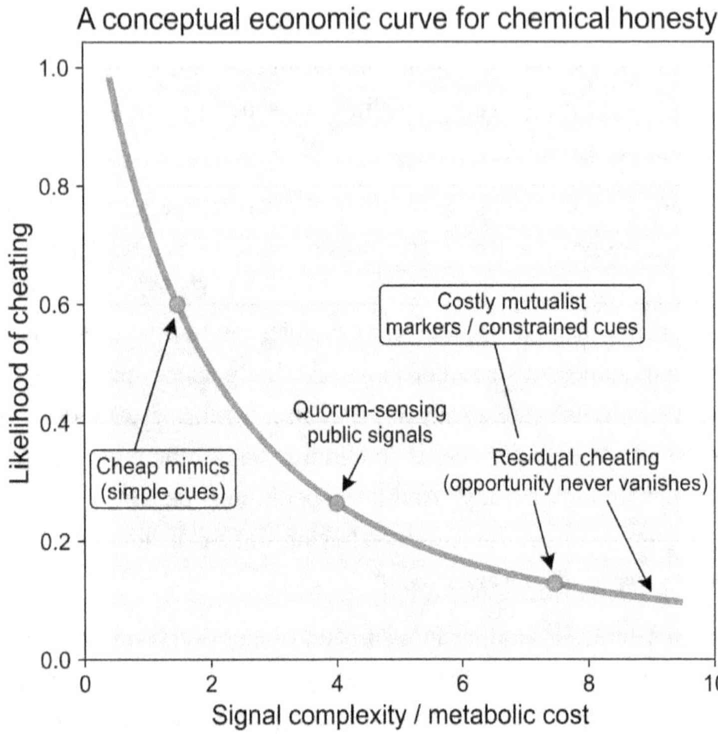

The most revealing aspect of this curve is not the downward slope but the stubborn floor. Even the most sophisticated recognition systems and the most expensive chemical words do not abolish deception. They merely make it rarer, riskier, or less profitable. That is enough. In evolution, stability is not the absence of conflict; it is the management of conflict so that cooperation can persist in the middle of temptation.

Seen this way, chemical trust is not a sentimental concept. It is an engineered outcome produced by molecules and by the selective pressures that shape them. A host learns, in the only way evolution can teach, to reward useful partners and to shut down the unhelpful.

A microbe learns that broadcasting cooperation can be profitable when the neighborhood is right and punishment is possible when it is not. A plant invests in chemistry that is expensive precisely because expense can make a message harder to counterfeit. The result is a world in which alliances can be forged without naïveté, with contracts written in gradients, receptors, and the quiet arithmetic of costs paid and benefits delivered.

Chapter 3

Nature's Stock Exchange

Beneath the forest floor, a ruthless economy operates with trading rules as strict as any human stock market. We investigate how fungi and plants haggle over phosphorus prices, employ sanctions against cheaters who take without giving, and maintain a fragile equilibrium through chemical currencies.

3.1 Supply, Demand, and Slime

Markets feel like a human invention: loud voices, flashing numbers, the nervous choreography of buying and selling. Yet the basic logic of a market is older than cities. Any time two parties can both gain by exchanging different things, and any time they can choose among partners, a kind of pricing system appears. Biology calls this *biological market theory*: cooperation can be understood as trade, shaped by scarcity, competition, and the ability to walk away.

The surprise is not that organisms trade. The surprise is how literal the trade can be. Sugars move in one direction, minerals in the other, and the exchange rate shifts with local conditions. Under a forest floor or in a farm field, a root can be a customer, a supplier, and a border checkpoint at once. The "market" is not a metaphor pasted onto nature for fun; it is a way of describing how countless microscopic decisions add up to stable cooperation without any goodwill required.

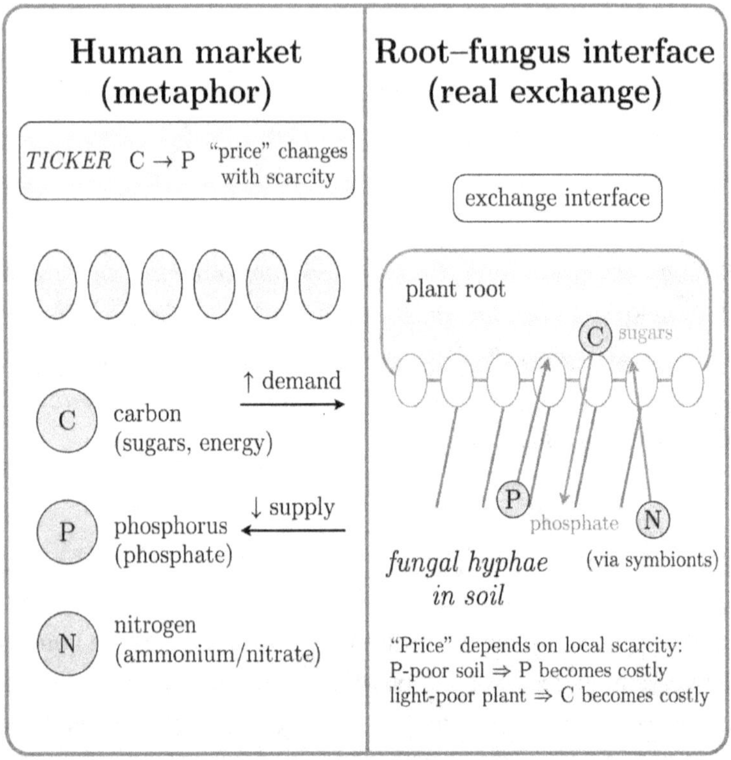

To make the idea concrete, it helps to name the currencies. The most important currency in terrestrial ecosystems is carbon, usually paid by plants as sugars or other carbon compounds produced by photosynthesis. Carbon is energy, but it is also the raw material for building bodies: cell walls, membranes, enzymes, spores. In many symbioses, carbon is the universal paycheck.

On the other side of the trade are nutrients that plants struggle to collect on their own, especially phosphorus and nitrogen. Nitrogen can be abundant in the atmosphere but stubbornly inaccessible; most plants cannot use N_2 gas directly. Phosphorus is plentiful in Earth's crust but often locked up in insoluble minerals or bound to soil particles. A plant root is a fine mining tool for some things, but not for all things. It is short, it is near-sighted, and it is constrained by physics: diffusion in soils is slow, and nutrients can be patchy. A fungus, by contrast, grows as threads that insinuate themselves into microscopic pores where roots cannot fit. The fungus is a living pipeline that turns distance into access.

Arbuscular mycorrhizal fungi, the most widespread root symbionts on the planet, are the archetypal currency exchangers. Their hyphae extend from the root into the soil, and their internal structures (arbuscules) push intimately into root cells, increasing the contact surface where molecules can cross from one partner to the other. The plant pays carbon; the fungus supplies phosphorus (and sometimes nitrogen and micronutrients). In the simplest telling, this sounds like a fixed bargain: a plant always gives sugar, a fungus always gives phosphate, and both always benefit.

The market view adds the missing variable: scarcity changes the terms. If a plant is growing in phosphorus-poor soil, phosphate is valuable. If phosphate is abundant, its value collapses. If the plant is shaded under a dense canopy, carbon becomes scarce; the plant's willingness to pay drops. Neither organism needs to calculate consciously. Physiology does the arithmetic. When carbon is plentiful, plants leak more carbon belowground, and fungi have more fuel to forage and transport. When carbon is tight, plants tighten the tap. When phosphorus is scarce, fungal foraging pays off, and fungal transport to the root increases the fungus's bargaining power. The "exchange rate" emerges from what each partner can credibly offer at that moment.

This is where the slime comes in. The underground economy is conducted through wet chemistry: thin films of water around soil particles, mucilage exuded by roots, and sticky fungal secretions that help bind soil into aggregates. These substances are not just plumbing; they are part of the trade. Root exudates include sugars, amino acids, organic acids, and signaling molecules. Some are deliberate invitations, chemical advertisements that say, in effect, *I am here, and I can pay.* Others are by-products, the inevitable leakage of a metabolically busy tissue. In soils, leaks are never neutral. They attract bacteria, fungi, and small animals, turning a root into a crowded street market with many potential traders and thieves.

Biological market theory, developed in the 1990s in part to explain cooperation in animal systems, becomes especially powerful in microbial worlds because microbes often have options. A flowering plant has many roots, and a single plant can host multiple fungal strains. A fungus can connect to multiple roots, sometimes even across different plant species. Partner choice does not have to mean choosing one partner for life; it can mean shifting investments among several. Even a small ability to bias exchange—to give slightly more carbon to one fungal partner than another—can create competition. Competition is what makes a market more than a swap. When multiple sellers offer phosphate, roots can be choosy. When multiple buyers offer carbon, fungi can be choosy. Price, in this setting, is not a number but a pattern: who gets rewarded, who gets ignored, and how strongly resources flow in each direction.

A concrete example begins with an everyday limitation: phosphorus is often the nutrient that throttles plant growth. Phosphate ions move slowly in soil water and are readily bound. A root hair can only explore a small neighborhood before the nearest phosphate becomes depleted. Fungal hyphae extend far beyond that neighborhood, essentially increasing the plant's reach. But the fungus is not a charitable extension cord. Hyphae cost carbon to build and maintain, and the fungus must decide where to invest its growth. If the fungus is connected to a plant that can afford to pay, expanding hyphae is worthwhile. If the plant is stingy, the fungus may invest elsewhere, store more phosphorus internally, or shift toward reproduction.

The intuition that resources flow both ways is easy; the evidence took time. One of the historical turning points came from agricultural chemistry long before anyone spoke of biological markets. In the nineteenth century, Justus von Liebig argued that crop yields are limited by the scarcest nutrient—the "law of the minimum." Farmers learned, sometimes painfully, that adding one nutrient can be useless if another is limiting. The global phosphate fertilizer industry grew out of this realization: phosphorus can be a hard ceiling on productivity. Yet there was always an awkward ecological question lurking beneath the fertilizers: if phosphorus is so limiting, how do plants persist in unfertilized soils over evolutionary time?

Part of the answer is that plants did not solve the problem alone. They hired contractors.

Imagine a young seedling in a phosphorus-poor patch of soil. It has a small photosynthetic budget and a large demand for phosphorus to build nucleic acids, membranes, and ATP, the energy currency of cells. A fungus in that soil faces the opposite: it can scavenge phosphorus effectively, but cannot photosynthesize. Their comparative advantages are complementary, and comparative advantage is the hidden engine of trade. In economics, it explains why a nation that is good at making everything still specializes. In biology, it explains why a plant that can absorb some phosphorus still outsources much of the work.

Once trade begins, a more subtle question follows: what stops cheating? A fungus could take carbon and deliver little phosphorus. A plant could accept phosphorus and then starve the fungus. Markets rely on enforcement, and the underground version relies on constraints built into physiology. The interface where the exchange happens is a controlled border. Plant membranes have transport proteins that gate phosphate entry. Fungi have their own mechanisms to package and move phosphorus, often storing it as polyphosphate chains inside their hyphae, like a warehouse inventory that can be shipped when the price is right. The very existence of storage changes negotiation: a fungus that can store phosphorus is not forced to sell immediately. It can wait.

Scarcity, in other words, can be felt as leverage.

The market metaphor becomes most vivid when conditions change. Consider a sunny meadow that gradually becomes shaded as shrubs grow. Photosynthesis slows. Carbon, the plant's payment currency, becomes more precious. A plant with less carbon to spare is less willing to subsidize fungal partners that provide only modest returns. At the same time, a shaded plant may still need phosphorus just as desperately as before, perhaps more so because growth is slower and each molecule invested must count. Now the plant faces a dilemma familiar to anyone with a tightening budget: what expenses are essential, and which contractors are overpriced?

The plant cannot renegotiate with words, but it can alter flows. Carbon delivery to the fungus can be reduced. Roots can shift architecture, producing more fine roots or longer hairs when mycorrhizal returns are low. Some plants, under certain conditions, down-regulate colonization by fungi. The point is not that plants always "win" these negotiations. A fungus is not a passive recipient. If carbon drops, the fungus may respond by reducing nutrient transfer, reallocating phosphorus to its own spores, or preferentially feeding those parts of the root system that still pay well. The exchange becomes a feedback loop: investment attracts investment.

This feedback is why supply and demand can be drawn in the mind even if it cannot be graphed easily. When phosphorus is scarce in soil, demand is high. Fungi capable of acquiring it become valuable partners. Plants may pay more carbon per unit phosphorus, because the alternative is stunted growth. When phosphorus is abundant, demand is low. Plants can afford to be stingier, and fungi must compete more intensely for carbon. The "price" of phosphorus in carbon terms falls.

Nitrogen produces a parallel story, but with a different cast. The most famous nitrogen traders are rhizobia, bacteria that live in root nodules of legumes and convert atmospheric nitrogen into ammonia. This is chemically expensive, like running an industrial factory inside a living cell. The plant pays with carbon and a sheltered home; the bacteria pays with usable nitrogen. Nitrogen fixation illustrates a crucial feature of biological markets: the commodity can be a service, not just a molecule. Rhizobia are not only delivering nitrogen; they are performing a reaction that the plant cannot.

The drama of services versus goods has human echoes. When industrial nitrogen fertilizer became widely available in the twentieth century, the price of biological nitrogen fixation effectively crashed for many crops. Farmers could buy nitrogen cheaply, and legumes became less central in some rotations. Yet the ecological cost of cheap nitrogen, from water pollution to greenhouse gas emissions, has forced a reappraisal. In a sense, human agriculture has been shopping in a market without paying the full environmental price, and the bill arrives elsewhere: in rivers, estuaries, and the atmosphere.

Underground markets have their own externalities, though they are usually internal to the ecosystem rather than exported. When fungi draw carbon from plants, that carbon is not available for leaves, seeds, or defensive chemicals. When plants divert carbon belowground, they may grow more slowly aboveground, changing competition with neighbors. Conversely, when fungal networks increase phosphorus uptake, plants may invest less in roots, changing soil structure and microbial communities. Every trade has opportunity costs. Cooperation does not erase conflict; it simply channels it into a negotiated form.

A vivid, almost cinematic example of scarcity shaping trade plays out after a disturbance. Picture a field after a fire or a plough. The soil surface is exposed, microbial communities are disrupted, and nutrients may be temporarily abundant due to ash or decomposition. In this pulse of availability, the relative advantage of fungal foraging can shrink. If phosphorus is suddenly easy to obtain, the plant's incentive to pay a fungal partner weakens. But pulses do not last. As nutrients are taken up or washed away, scarcity returns. Fungal networks, if they persist through the disturbance, can buffer plants against the crash by maintaining access to patchy reserves. In a market sense, fungi can function as both suppliers and insurers.

Insurance introduces another subtlety: the value of a partner is not only what it provides today, but what it can provide under stress tomorrow. A fungal species that is mediocre in rich soil may be essential in drought. A bacterial partner that grows slowly may be unusually reliable under fluctuating conditions. Markets in nature are not optimized for quarterly profits; they are shaped by survival and reproduction across variable environments. That difference matters when borrowing economic language. Words like "rational" can mislead if they imply conscious calculation. The rationality here is evolutionary: lineages that allocate resources in ways that tend to increase their reproductive success persist, and those patterns can look eerily like strategy.

The molecular form of the currencies also matters, because chemistry constrains what can be traded and how quickly. Carbon can be transferred as sugars such as glucose and sucrose, but also as lipids and other compounds, depending on the symbiosis. Phosphorus is often transferred as inorganic phosphate, but within fungi it is frequently transported and stored as polyphosphate, a polymer that can be shuttled along hyphae like freight in a train of linked cars. Nitrogen may move as ammonium, nitrate, or amino acids, and the chemical form affects leakage, theft, and control. A molecule that diffuses readily is easy to steal. A molecule that requires a membrane transporter is easier to regulate. Many symbioses, unsurprisingly, rely on forms that can be gated at membranes, turning chemistry into contract.

Contracts, however, are never perfect. Soil is crowded, and every resource gradient attracts opportunists. Carbon-rich exudates feed not only helpful microbes but pathogens. Phosphate released near roots can be intercepted by other organisms. This is one reason mycorrhizal fungi deliver phosphorus directly at intimate interfaces rather than dumping it into soil water. Direct transfer reduces interception, like paying with encrypted bank transfers instead of leaving cash on a park bench.

Even the best-enforced trades operate on averages. A plant can bias carbon toward good partners, but cannot always discriminate perfectly among fungal strains, especially when hyphae are interwoven. A fungus can preferentially allocate nutrients to roots that pay well, but roots are not isolated accounts; they are parts of a whole organism with shared circulation. In economic terms, there are transaction costs and information limits. In biological terms, there are constraints of anatomy and diffusion.

Recognizing those constraints prevents the market metaphor from becoming too neat. Not every symbiosis behaves like a competitive exchange with fluid pricing. Some partnerships are more like long-term alliances with limited choice, where the "rate" is relatively fixed because alternatives are rare or because the relationship has become developmentally integrated. Some exchanges may be driven less by partner choice and more by internal sink strength: whichever partner can use a resource fastest draws it in, not because it is rewarded, but because it creates the strongest pull. In practice, both forces can operate together. A plant may provide more carbon to a fungus because that fungus delivers more phosphorus (a reward), and also because the plant tissue colonized by that fungus is growing faster and therefore pulls more carbon (a sink effect). Biology is comfortable with mixed motives, because it has no motive at all—only mechanisms.

Still, the market perspective does something valuable: it encourages a particular kind of question. Instead of asking whether a symbiosis is "good" or "bad," it asks what is scarce, who controls access, what information can be sensed, and what alternatives exist. Those questions can be answered with experiments. Alter light levels and watch carbon allocation shift. Enrich soil with phosphate and see whether fungi reduce delivery or plants reduce payment. Introduce multiple fungal partners and measure whether the plant biases its investments. The market metaphor invites measurement, not moralizing.

Underfoot, the exchanges are incessant. Carbon trickles down as sugars and other compounds, feeding not only the obvious partners but whole constellations of microbes. Phosphorus and nitrogen move upward in pulses and streams, sometimes delivered directly across membranes, sometimes processed through intermediate traders. The soil becomes a living ledger written in isotopes and transport proteins, in sticky films of water and the branching geometry of hyphae. Cooperation emerges not because nature is kind, but because scarcity makes collaboration profitable, and because chemistry makes profit enforceable.

3.2 The Art of the Deal

A market without rules collapses into theft. The same is true underground. If a plant is going to pour hard-won carbon into the soil, and if microbes are going to invest scarce minerals into a root, both need some assurance that the other side will not simply take and run. Nature cannot call a lawyer. It builds enforcement into anatomy, timing, and chemistry, with consequences that can look uncannily like screening, auditing, and penalties.

Begin with a simple problem: a plant root is not negotiating with a single, well-mannered partner. It is surrounded. The soil around roots, the rhizosphere, is a crowded zone where bacteria, fungi, and tiny animals respond within minutes to any leak of sugar or amino acids. Many of those organisms are harmless freeloaders. Some are pathogens. Some are potential allies. A plant that treats them all the same becomes an open bar. A plant that can distinguish among them—even roughly, even probabilistically—has an advantage. Biologists call that ability *partner choice* when it happens before or during the start of a relationship, and *sanctions* when it happens after a partner has already moved in and is underperforming.

The vocabulary is economic, but the machinery is physical. Roots control entrances. They decide where to build nodules, where to permit a fungus to cross the outer cell layers, where to maintain intimate interfaces, and where to let those interfaces degenerate. Those decisions are made with signals: molecules exchanged early, long before the main currencies start to flow. This is screening in the most literal sense. A plant's surface is a test.

A classic example comes from legumes and their nitrogen-fixing bacteria, rhizobia. Nitrogen fixation is expensive: converting atmospheric N_2 into ammonia requires energy and a delicate enzyme complex that is poisoned by oxygen. Rhizobia solve the oxygen problem by living inside root nodules that provide a low-oxygen environment while still allowing respiration. The plant pays heavily to build these nodules, supplying carbon and maintaining a specialized tissue. That investment creates a temptation. Any bacterium that can enter a nodule and enjoy the carbon without paying nitrogen would be, in the narrowest sense, a successful cheat.

Plants are not helpless. One of the most elegant controls is simply the ability to decide who gets a room in the hotel. A legume does not form nodules with just any soil bacterium; it responds to particular chemical signals (so-called Nod factors) produced by compatible rhizobia. That specificity is already a gate. But gates are never perfect, and evolution is relentless. The more interesting protection comes after the bacteria are inside.

Some of the best evidence that plants can punish poor performers comes from experiments in which rhizobia strains differ in how much nitrogen they fix. When plants are infected with a mixture, nodules containing ineffective strains tend to receive fewer resources and can be physiologically constrained. The details vary among legume species and experimental systems, but the logic is consistent: the plant changes the internal environment so that a nodule that does not deliver nitrogen becomes a worse place to live. A useful way to picture this is as a local budget. The plant does not have to identify the genetic identity of every bacterium; it only needs to monitor output at the level of a nodule. If ammonia or related nitrogen products are low, carbon flow can be reduced, oxygen supply can be restricted, or the nodule can be allowed to senesce early. The unit of enforcement is not the individual microbe but the compartment.

Compartmentalization shows up again and again in biological contracts. A fungus does not tap directly into a plant's bloodstream. In arbuscular mycorrhizal symbiosis, for example, fungal arbuscules form inside root cortical cells, but they remain separated from the plant cytoplasm by a plant-derived membrane. That extra membrane is more than a kindness of evolution; it is an interface where transporters can be regulated. The plant can upregulate phosphate transporters when the fungus is delivering, and downregulate when the fungus is not. The fungus can adjust its own transport processes. Even if neither partner has a “choice” in the human sense, both have levers.

The insight that partners can actively reward and punish is relatively young, because it required the ability to measure exchange directly. For a long time mycorrhizal fungi were described in almost pastoral terms: fungi help plants, plants feed fungi, everyone benefits. That story was persuasive but vague. The breakthrough came from turning nutrients into traceable currencies.

Stable isotope tracers made the underground economy legible. If carbon is labeled with ^{13}C (a non-radioactive heavier isotope), its movement into fungal tissues can be tracked. If phosphorus is labeled with ^{33}P or nitrogen with ^{15}N, their transfer can be quantified. What had been metaphor became measurable flow. With tracers, it became possible to ask an almost impolite question: does a plant actually pay more to a fungus that provides more? And does a fungus actually deliver more to a root that pays more?

Experiments by Toby Kiers and colleagues around 2011 made this reciprocity vivid for mycorrhizas. By manipulating which parts of a root system received which fungal partners and by tracking labeled nutrients, they showed that plants can preferentially allocate carbon to fungal partners that deliver more nutrients, and that fungi can bias nutrient delivery toward roots providing more carbon. The effect is not a tidy one-to-one swap, and it does not require conscious accounting. It is enough that both partners can detect local conditions at the interface and adjust transport accordingly. The remarkable point is not that cooperation exists; it is that it is actively negotiated.

Negotiation is where the "art" lies, because perfect enforcement is rarely possible. Think about the informational challenge. A plant cannot easily look at a fungal hypha and ask how hard it worked to obtain phosphate. It can only measure what arrives. A fungus cannot easily assess how much photosynthesis is possible for the plant overall, only what carbon is supplied locally. These constraints create opportunities for misunderstanding and exploitation. Nature's response is often redundant safeguards: screening before entry, monitoring during exchange, and penalties that are costly but less costly than being cheated indefinitely.

Screening starts early and can be surprisingly sophisticated. Many symbioses begin with chemical courtship. A root under nutrient stress may release compounds that increase fungal branching or attract beneficial bacteria, effectively advertising a willingness to trade. Microbes, in turn, release signals that alter root development, sometimes triggering specialized structures. These cues can be honest or deceptive. A pathogen can mimic friendly signals. A would-be mutualist can exaggerate. Plants therefore benefit from not relying on a single cue. They can demand a sequence of confirmations: signal A, then signal B, then actual delivery. This resembles a background check more than a handshake.

A concrete human-scale analogy helps: when you hire a contractor, you might look at a portfolio (signals), check references (correlated cues), start with a small job (limited exposure), and only then approve a larger project (major investment). Plants cannot browse portfolios, but they can do something similar by staging commitments. Early interactions may involve small exudate payments that are cheap to lose. Only after a fungus penetrates and forms arbuscules does the plant open major carbon flows. The temporal structure of the relationship is itself a screening device.

Even then, plants face the problem of mixed infections. A single root system can host multiple fungal genotypes. A legume can harbor different rhizobial strains across different nodules. If a plant pays at the level of the whole organism, cheaters can hide among cooperators, taking advantage of the plant's global generosity. Effective sanctions often require local accounting: rewarding the good node, penalizing the bad one. This local accounting is easier when symbionts are spatially segregated. A nodule is a discrete unit. An arbuscule is a discrete unit. A leaf gall or a coral polyp can be a discrete unit. Nature repeatedly solves contract enforcement by building many small compartments rather than one big shared pool.

The other side of enforcement is partner choice in the presence of competition. Biological market theory predicts that if multiple potential partners exist, the ability to choose can maintain cooperation even if sanctions are imperfect. This can feel counterintuitive. If cheating is possible, why doesn't cheating spread? One answer is that cheating can be locally successful but globally punished by market dynamics. A fungus that withholds nutrients may gain carbon from one host, but if hosts in general bias carbon toward more generous fungi, the stingy strain loses out over time. Similarly, a rhizobial strain that fixes little nitrogen might still form nodules, but if plants tend to allocate less carbon to those nodules, the strain's long-term reproduction suffers.

This is not a moral tale; it is a story about differential growth and survival. In microbial worlds, small differences in resource flow compound quickly. A slight reduction in carbon can mean fewer spores, slower hyphal expansion, fewer opportunities to colonize new roots. Selection can amplify the effect of tiny sanctions.

Sanctions can also be indirect, and that subtlety matters because it complicates the popular image of a host "punishing" a symbiont like a referee issuing a card. Sometimes the host is not punishing; it is reallocating. If a plant's phosphate transporter is not engaged because phosphate is not arriving, the arbuscule may degenerate. The fungus experiences that as loss of access, but the plant's action may be simply physiological housekeeping: maintain interfaces that are productive, dismantle those that are not. In other words, enforcement can be an emergent consequence of efficiency.

That interpretation sits at the heart of a real debate in mycorrhizal research. Some scientists emphasize reciprocal rewards and partner choice as central to mycorrhizal stability; others stress that resource exchange can be driven largely by sink strength, functional differences among fungi, and constraints on discrimination. Both perspectives can be right depending on the species involved and the environment. A plant might be capable of choosing among fungal partners in one context and effectively stuck in another. A fungus might have leverage in phosphorus-poor soils but little leverage in phosphorus-rich soils. The danger is not disagreement; it is taking one mechanism and treating it as a universal law.

A historical anecdote from economics helps keep the biology honest. The idea that self-interest can produce order without central control is often associated with Adam Smith's "invisible hand," published in 1776. What is less often recalled is how much Smith worried about monopoly and collusion: markets do not automatically produce fairness or even efficiency. They work under certain conditions—especially competition, transparency, and the ability to exit. Nature's markets face analogous constraints. If there is only one viable fungal partner in a soil, the plant's bargaining power is low. If a plant is so dependent on a symbiont that it cannot survive without it, exit is impossible. If information about partner quality is noisy or delayed, cheaters can prosper. Contract enforcement in biology is therefore not a victory lap for market metaphors; it is a way of stating precisely what must be true for cooperation to be stable.

One of the most emotionally resonant places to see these principles is in agriculture, where humans unwittingly become part of the contract. Consider a farmer planting legumes. The field contains rhizobia of varying quality. The plant's sanctions can help, but humans can change the incentives by adding nitrogen fertilizer. When nitrogen is abundant in the soil, the plant needs the bacteria less. It may form fewer nodules or supply them less. From the bacteria's point of view, the market for nitrogen collapses. This can feed back into evolution: if fixation is no longer rewarded, strains that invest less in fixing may outcompete better fixers, because fixing is costly. Then, if fertilizer inputs later decrease, the farmer discovers that the biological supply chain has degraded. The field has "learned" a new market reality.

A similar dynamic can occur with phosphorus and mycorrhizas. Heavy phosphate fertilization can reduce the plant's dependence on fungal partners, changing colonization and investment patterns. The consequences can be subtle: shifts in fungal community composition, altered soil aggregation, different disease susceptibility. The main point is that sanctions and partner choice are not abstract curiosities; they are mechanisms by which ecosystems maintain cooperation, and they can be disrupted when humans alter the currency landscape.

The most striking feature of biological sanctions is their quietness. There is no dramatic confrontation, no single moment where a cheater is exposed. Instead there is a throttling: less sugar through a membrane transporter, less oxygen diffusing into a nodule, fewer lipids delivered to a hyphal interface. Life is built on gradients, and sanctions are often gradient changes.

That quietness is easy to miss, because it does not match the way humans imagine punishment. Yet it may be closer to how most real-world enforcement works even in human systems. Credit becomes harder to obtain. Contracts are not renewed. Access is restricted. The penalty is deprivation of future opportunity rather than a direct blow. In the soil, future opportunity is everything. A microbe without carbon cannot maintain its enzymes. A fungus without host access cannot expand. A nodule starved of oxygen cannot power nitrogen fixation even if it “wants” to.

If all this seems to risk turning nature into a caricature of Wall Street, the corrective is chemistry. The negotiations are conducted through molecules that obey physical laws. A transporter protein has a maximum rate. A membrane has limited surface area. A root cell can host only so many arbuscules before it becomes crowded and stressed. A plant can allocate only so much carbon belowground before leaves suffer. These constraints impose discipline. They ensure that cooperation is never free and that enforcement never comes without cost.

The elegance of the system is that the costs can be paid locally. A plant does not have to shut down its entire relationship with fungi to penalize one unproductive interface; it can prune. A fungus does not have to abandon an entire plant to respond to poor payment in one root region; it can redirect. Local control turns a potentially brittle partnership into something adjustable, like a set of valves rather than a single pipeline.

In the end, the “deal” is not a promise; it is a moving equilibrium maintained by continual measurement. Sugars arriving? Keep the arbuscule alive. Phosphate arriving? Keep the carbon flowing. Nitrogen arriving? Keep the nodule oxygenated. The contract is written in flux, and enforcement is simply the ability to alter that flux in response to performance. In a world where every organism is under pressure to survive, that is about as close to trust as chemistry can get.

3.3 Too Big to Fail

A single trade between a root and a fungus is easy to picture: sugar for phosphate, a handshake made of membranes and transport proteins. Scale that up to a meadow or a forest, and the scene changes. The underground economy becomes a network, not a pairwise bargain. One fungus may connect to many plants; one plant may host many fungi; bacteria and animals graze, prune, and redirect the flow. At that point, the most important question is no longer "Is this partnership fair?" but "What happens when parts of the system fail?"

The phrase *too big to fail* belongs to finance, but the underlying fear is ancient: when a system becomes tightly interconnected, local trouble can spread. In banking, an institution's collapse can freeze credit. In ecosystems, the loss of a few key symbionts can freeze nutrient flows, weaken plants, and reshape entire communities. The twist is that ecological networks can also do the opposite. The same dense connectivity that allows contagion can provide redundancy, alternative routes, and a kind of resilience that no isolated partnership could achieve.

One way to see the difference is to picture two styles of underground trading. In one, each plant has its own specialist fungal partner, and each fungus works with a narrow set of hosts. The network is sparse. If a specialist disappears, its host is stranded. In the other, a handful of generalist fungi act like hubs, interacting with many plants, while plants themselves often associate with several fungal species. The network is dense and typically *nested*: specialists tend to interact with generalists rather than with other specialists. Nestedness matters because it creates a safety net. A rare plant species can still plug into the system through common fungi; a rare fungus can still find carbon through common plants.

This is where market logic meets network logic. Markets explain why a plant might reward a better partner and punish a worse one. Networks explain why those individual choices can stabilize, or destabilize, entire communities. A plant can be choosy and still be embedded in a web where its choices have spillover effects. If a plant diverts carbon toward one fungal partner, it may indirectly deprive other plants that share that fungus of some nutrient flow, or it may enrich them if the fungus responds by expanding its foraging. Cooperation can become a public good, or a private advantage, depending on network structure.

A familiar idea in ecology is that diversity stabilizes function: having many species can buffer an ecosystem against shocks because different species perform similar roles under different conditions. Mycorrhizal fungi are a textbook case of this principle, but with a chemical twist. Fungi are not interchangeable pipes. They differ in how they explore soil, how they store nutrients, which enzymes they deploy, how they tolerate drought, heat, acidity, salinity, and heavy metals, and how strongly they *demand* carbon. Some are good at scavenging phosphate from tightly bound pools; others are better at rapid uptake from pulses. Some build extensive networks that can bridge gaps; others concentrate on tight, local foraging. When fungal diversity drops, the ecosystem does not merely lose species; it loses functional options for moving atoms through the landscape.

A concrete illustration came from work in the 1990s showing that altering arbuscular mycorrhizal fungal diversity could change plant community composition and productivity. The headline message was not that fungi help plants (everyone already suspected that), but that *which* fungi are present can change *which* plants win. In a mixed plant community, small differences in nutrient acquisition become competitive advantages. A fungus that benefits grasses more than forbs can tip the balance toward grass dominance; a fungus that favors certain legumes can change nitrogen inputs, with cascading consequences for neighbors. The underground network becomes a hidden hand that shapes aboveground diversity.

These effects can be dramatic in stressful environments where small nutritional edges matter most. In nutrient-poor tropical soils, phosphorus can be so limiting that access to the right fungal partners makes the difference between persistence and failure. In cold or dry systems where root growth is constrained, fungal exploration can substitute for root expansion. In these contexts, losing key fungal lineages is less like losing a minor service and more like losing infrastructure.

Infrastructure is an apt word because it captures two features at once: the network is costly to build, and many organisms rely on it. Hyphae require carbon. Soil aggregation built by fungal threads and secretions changes water retention and aeration. Fungal networks influence how quickly leaf litter decomposes and how nutrients are retained versus leached away. This is why mycorrhizal fungi are sometimes called ecosystem engineers: they reshape the environment in ways that feed back into who can live there.

Yet infrastructure also implies vulnerability. Roads can be washed out. Power grids can be attacked. Underground networks can be broken by ploughing, soil compaction, mining, erosion, and changes in vegetation. The most obvious consequence is that plants may grow less well. The more subtle consequence is that the *pattern* of growth changes: which species establish after disturbance, how quickly forests regenerate, how grasslands respond to drought, how readily invasive plants take hold.

Common mycorrhizal networks (often abbreviated CMNs) are the most evocative example of scaling up from pairwise trades. A CMN exists when a fungal mycelium connects the roots of multiple plants, potentially allowing resources and signals to move between them.

This idea has a long and sometimes contentious history because it is easy to romanticize. The image of a benevolent "wood-wide web" that shares food evenly among forest members is emotionally satisfying, but nature is not a social welfare program. Transfers through networks can happen, but they can be conditional, asymmetric, and shaped by competition.

The chemistry makes that clear. Carbon does not flow through a CMN because plants are generous; it flows because concentration gradients, sink strength, and fungal allocation rules permit it. Nutrients can move from one plant to another if the fungus routes them that way, which may occur when one plant is a strong carbon source and another is a strong nutrient sink, or when the fungus benefits by keeping multiple hosts alive. Signals can move too: defensive cues, stress hormones, or microbial metabolites that prime neighboring plants. But even signaling can be self-serving. A fungus that warns a host of impending herbivory may be protecting its own carbon supply.

A helpful way to think about this is to imagine that the fungus is not a neutral wire but an active broker. In a financial network, money can be routed through institutions that profit from the routing. In a CMN, the fungus sets terms through its growth patterns, storage, and local exchange rates. That brokerage can stabilize the system by smoothing out fluctuations, or it can amplify inequality by channeling resources toward particular hosts that pay well.

This is where earlier ideas about sanctions and partner choice take on a new meaning. If a plant can penalize an unproductive fungal partner locally, and a fungus can shift nutrients toward better-paying roots, then a shared network becomes a dynamic arena of local negotiations. The emergent pattern is not necessarily fairness; it is persistence. A network can persist even when some individual interactions are exploitative, provided that, on average, enough reciprocal exchanges occur to keep the main lines of trade open.

Network resilience depends on architecture. Ecologists use terms like *connectance* (how many links exist), *nestedness* (whether specialists link to generalists), and *modularity* (whether the network splits into semi-independent clusters). These are not abstract graph properties; they translate into ecological outcomes. Highly modular networks can prevent disturbances from spreading, like firebreaks. Highly connected networks can provide alternative routes when one link fails, like detours around a closed bridge. Nestedness can protect rare species by giving them access to robust hubs.

The same logic explains why some species function as hubs in ecological networks and why losing them can have outsized effects. A generalist mycorrhizal fungus that associates with many plant species can act as an underground keystone: not because it is noble, but because many trades pass through it. If that hub is lost, the network does not simply lose one species; it loses a disproportionate fraction of interactions, and the remaining system can fragment into isolated components. Fragmentation is a kind of market failure. Trades that were once possible become impossible, not because the currencies are gone, but because the exchange venues disappear.

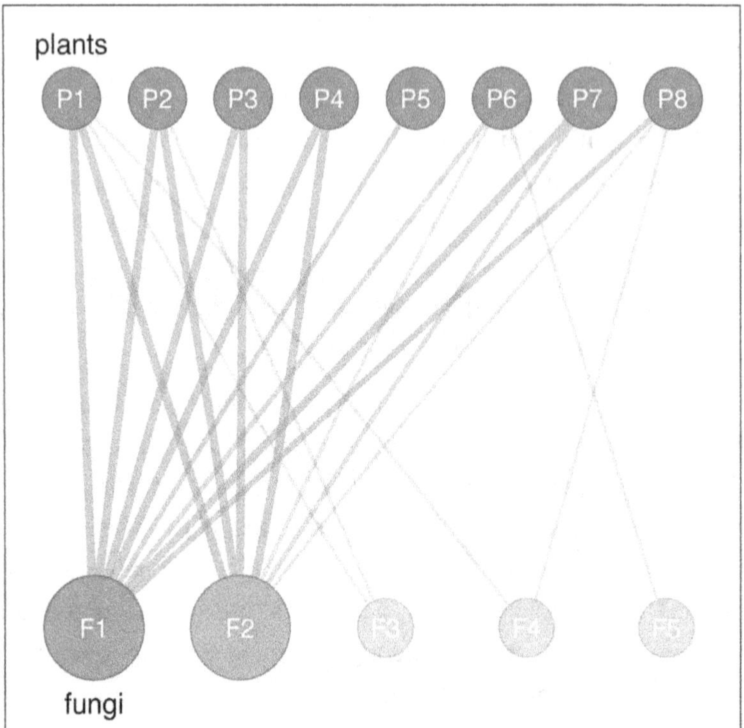

generalist hubs keep the network connected

The picture above is not a real forest, but it captures a crucial intuition: hubs reduce the chance that any one plant will be isolated. That intuition can be tested in the real world by looking at what happens when ecosystems are simplified. Monoculture crops, for example, often rely on a narrower set of microbial partners than diverse natural communities. That simplification can be efficient in the short term, but it can also reduce the buffering capacity that comes from redundancy. When a stress arrives that the dominant partners cannot handle, there are fewer alternatives waiting in the wings.

The strongest stresses are increasingly anthropogenic. Climate extremes alter the relative value of water, nitrogen, and phosphorus.

Land-use change can remove host plants needed by certain fungal lineages. Pollution and excess nutrients can distort exchange rates, weakening the incentives that normally maintain cooperation. Under heavy nitrogen deposition, for instance, some plants may invest less in mycorrhizas because nitrogen is no longer limiting, even though fungi may still be important for phosphorus or drought tolerance. Relationships can slide along the mutualism—parasitism continuum: a partner that was once valuable becomes a cost under new conditions, and the network rewires.

Rewiring is not always possible. Many mycorrhizal fungi disperse slowly and may not recolonize disturbed soils quickly. Inoculation can help in restoration projects, but it is tricky. Introducing the wrong fungi can be ineffective or even harmful, and moving microbes across regions carries risks. The underground infrastructure is local and historically contingent; it reflects thousands of years of climate, vegetation, and soil development.

A recent mapping effort made the conservation stakes uncomfortably explicit. A large machine-learning analysis led by researchers associated with the Society for the Protection of Underground Networks and published in *Nature* in 2025 used tens of thousands of geolocated soil samples to predict global patterns of arbuscular and ectomycorrhizal fungal richness. One result landed like a warning label: less than 10% of predicted mycorrhizal fungal richness hotspots appear to fall within protected areas, with an overall estimate around 9.5% for richness hotspots. This is a striking mismatch between what ecosystems rely on and what conservation policy typically protects. Protected areas are often designed around charismatic plants and animals, or around scenic landscapes, while the biodiversity that makes soil work has remained largely invisible.

Visibility matters because protection is partly a legal and cultural act. People fight for what they can picture. A tiger has a face; a fungus has threads. A forest has grandeur; a soil community looks like dirt. The irony is that mycorrhizal fungi are not marginal. They are deeply implicated in carbon storage, nutrient retention, and plant health. When conservation overlooks them, it risks protecting the roof while letting the foundation crumble.

There is also a more intimate reason to care. The stability of these networks is not only about wilderness. It is about food. Many crops depend on mycorrhizas to some degree, especially in low-input systems. Even in high-input agriculture, where fertilizers can substitute for some nutrient acquisition, mycorrhizal networks can influence soil structure, drought resilience, and disease dynamics. The more erratic the climate becomes, the more valuable these belowground stabilizers may be. Resilience is often what remains after efficiency has failed.

Network thinking also changes how one interprets "cheaters." In a pairwise relationship, cheating seems like a simple moral category: one partner takes without paying. In a network, the line blurs. A fungus that takes carbon from one plant and delivers nutrients preferentially to another might look like a cheat from the first plant's perspective, but from the fungus's perspective it may be maintaining multiple hosts, ensuring continuity of carbon supply over time. A shaded seedling receiving carbon via a network may look like a freeloader, but keeping that seedling alive could maintain future host diversity for the fungus or stabilize the plant community. Some transfers that look like generosity may be long-term investment by the broker.

That does not mean networks are benevolent. It means that the unit of "benefit" can shift. Selection does not optimize the welfare of a plant individual or a fungal individual; it shapes strategies that succeed in the contexts organisms typically face. Those contexts include disturbance, competition, and variability. Networks can be an evolved response to unpredictability, spreading risk across multiple partners the way diversified portfolios spread financial risk.

And yet portfolios can also be fragile if everyone holds the same assets. Ecological analogues exist. If many plants depend on the same dominant fungal hubs, then the hubs become critical points of failure. If environmental change disproportionately harms those hubs, the network can collapse rapidly. Diversity in hubs, not just diversity in peripheral species, becomes important. The safest underground economy is not one with a single giant bank, but one with multiple institutions that can substitute for each other.

Under the surface, the most dramatic events are often quiet: a thinning of hyphae after repeated tillage, a shift in fungal composition after years of fertilization, a fragmentation of networks after compaction, a slow loss of rare lineages as landscapes are converted. Aboveground, these quiet events can eventually become visible as poorer regeneration, reduced drought tolerance, and simplified plant communities that support fewer insects and birds. The causal chain is long, but the chemistry at its start is immediate: less phosphorus arriving at membranes, less carbon invested in roots, fewer exchanges closing the nutrient cycles that make a landscape fertile.

Soil, in the end, is not merely a medium that holds plants upright. It is a living marketplace whose stability depends on networks of exchange. When those networks are intact, plants can survive lean seasons, colonize harsh ground, and assemble diverse communities from scarce resources. When they are severed, the same landscape can look intact from a distance yet behave like an economy in slow recession: nutrients leak away, resilience thins, and recovery from shocks takes longer each time.

Chapter 4

The Art of Deception

Nature is often romanticized as a harmonious web, but beneath the surface lies a ruthless marketplace where fraud is a constant temptation. This chapter peels back the layers of cooperation to reveal the chemical disguises used by freeloaders and the physiological policing strategies that prevent these fragile alliances from collapsing into exploitation.

4.1 Counterfeiters in the Crowd

A mutualism runs on trust, but not the sentimental kind. It is closer to a well-designed market: each partner offers something costly to produce and difficult to obtain alone, and each partner has ways of checking that the trade is real. When a plant feeds carbon to a fungus, or a flower advertises to a pollinator, chemistry supplies the receipts. Tiny molecules act as infochemicals: information-bearing compounds that announce identity, readiness, and value. In principle, this should make cooperation robust. In practice, it creates a target.

Wherever a biological interaction depends on a recognizable signal, evolution discovers a familiar trick: counterfeit the signal, collect the payment, skip the work. A parasite does not need to overpower its host if it can persuade the host to open the door. A free rider does not need to dismantle a mutualism if it can ride inside it, extracting benefits that were meant for honest partners. The result is less like a clean break between “mutualist” and “parasite” and more like a crowded party where most guests are genuine, a few are helpful, and some have learned to imitate the host’s handshake.

Chemistry makes this possible because living systems often rely on *pattern recognition*. A receptor does not read a full autobiography; it binds a molecule with a particular shape, charge, and timing. Many cooperative deals begin with a small set of cues that have to be reliable enough to guide behavior quickly: “I am the right partner,” “I am fertile,” “I am a nestmate,” “I am fixing nitrogen,” “I am not a threat.” These cues are often called “keys” because they fit molecular “locks” on cell surfaces. But a lock that opens for a friend can also open for a skilled forger.

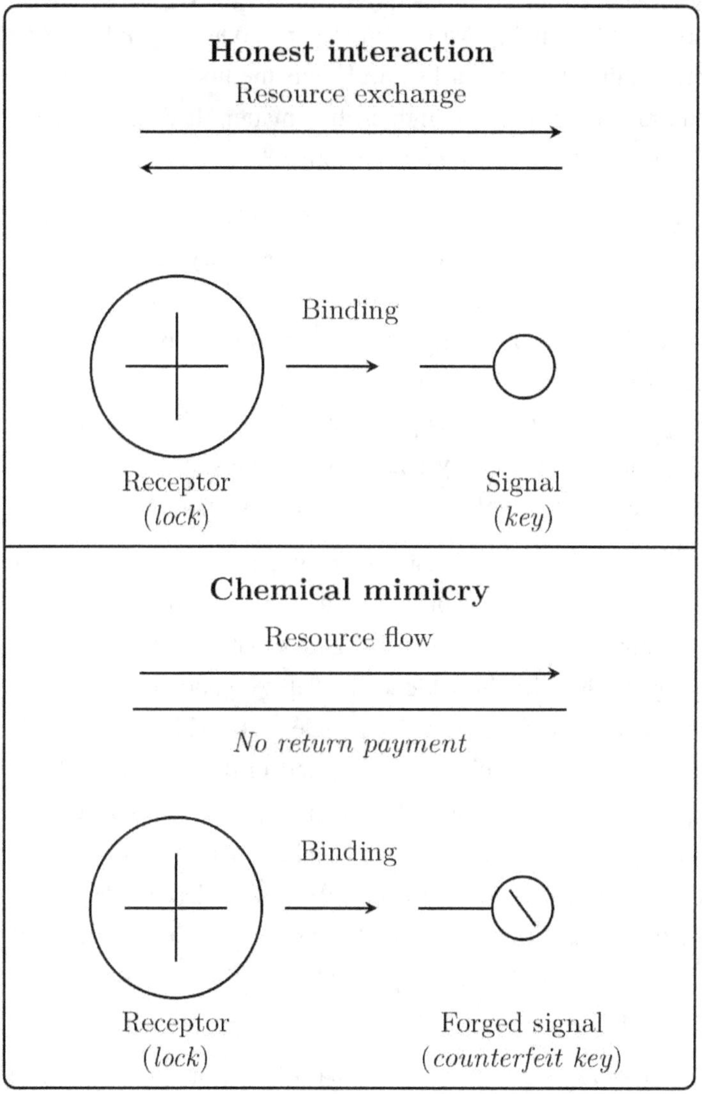

The appeal of counterfeiting is energetic. Honest signals are often expensive or tightly coupled to real performance. A flower that produces nectar invests sugars; a bacterium that fixes nitrogen invests enzymes and energy; a fungus that delivers phosphorus invests transporters and hyphal growth. Mimicry, by contrast, can be cheap if it only requires the outward sign. This is the first reason deception is common: information is lighter than matter. It is usually easier to *look* cooperative than to *be* cooperative.

A second reason is that many signals evolved for speed rather than security. A legume root cannot interview every microbe in the soil. A social insect colony cannot run genetic tests at the nest entrance. A male insect responding to a sex pheromone is built to be fast and decisive; hesitation is costly in mating. Evolution often favors a recognition system that is good enough most of the time, because the environment is noisy and decisions must be made under time pressure. The price of a quick heuristic is vulnerability.

One of the most vivid examples of chemical counterfeiting is the "sexual swindle" of *Ophrys* orchids. A male bee or wasp is not shopping for nectar; he is searching for a mate. Many *Ophrys* species exploit this by releasing a blend of volatile compounds that resembles the sex pheromone of receptive females. The male arrives, performs pseudocopulation with the flower, and in the process collects and deposits packets of pollen. The orchid receives the service of pollination; the insect receives no mate and often no food reward. Florian Schiestl and Manfred Ayasse (1999) showed that the deception can be extraordinarily specific: an *Ophrys* species may mimic the chemical proportions of a single pollinator species so closely that other insects ignore it.

The trick works because the pollinator's sensory system is tuned to a narrow chemical channel, almost like a radio set to a particular frequency. Natural selection has already done the orchid's advertising research: it knows what message the male cannot resist. The orchid is not merely imitating a generic perfume; it is forging a private passcode, the sort of signal that is usually hard to fake because it is tied to reproductive physiology in the female insect. But the orchid does not need to grow ovaries to produce a hydrocarbon blend. The message is enough.

If this feels like a con, it is worth pausing on the insect's perspective, because it shows how deception can persist without "stupidity" on either side. A male that refused to respond strongly to female pheromones would miss genuine mates. The cost of being cautious can exceed the cost of occasionally being duped, especially if the orchid is rare relative to females. The orchid, for its part, benefits from persuading the male to attempt copulation, but it cannot push the system too far: if deception became so common that males learned to avoid those scents, the strategy would collapse. The success of the swindle depends on frequency, ecology, and the limits of learning.

Not all counterfeiting is so theatrical. Many cheats work by taking advantage of mutualisms that operate under the surface, where the currency is not pollen but nutrients and protection. Consider the bustling trade between plants and mycorrhizal fungi, where plants supply carbon and fungi supply minerals such as phosphorus. The handshake between partners involves both metabolites (food) and infochemicals (signals of compatibility). The details can be highly specific: particular plant compounds can stimulate fungal growth and branching, and fungal signals can influence plant immune responses so that the fungus is tolerated rather than treated as an invader.

This creates a tempting opportunity for a fungal free rider: persuade the plant that a productive partner has arrived, then collect carbon while delivering little phosphorus. The challenge is that underground, "delivery" is hard to audit directly. A plant cannot watch each fungal hypha unloading each phosphate ion. Instead, it must rely on physiological proxies: growth responses, nutrient status, local concentration gradients, patterns of transporter activation. Those proxies can be manipulated. A fungus that is good at triggering the plant's cooperative physiology may receive carbon even if it is not, in net terms, a good trade partner.

At the molecular level, this kind of exploitation can look like a carefully chosen lie: a compound that resembles a legitimate fungal signal binds to a plant receptor and suppresses defenses or encourages carbon flow. The story is not always villain versus victim; fungi vary in benefit, and the environment changes what counts as "good service." A fungus that offers little phosphorus in a rich soil may still be valuable in drought, or as protection against pathogens. But the possibility of exploitation is built into the system whenever carbon allocation is influenced by signals that can, in principle, be mimicked.

If orchids show how to counterfeit desire, social insects show how to counterfeit belonging. Ant colonies are among the most chemically policed societies on Earth. Individuals recognize nestmates largely through blends of cuticular hydrocarbons: waxy molecules on the body surface that function as a colony odor. Think of them as a living uniform. Guards at the nest entrance sample the chemical profile of incoming ants with their antennae. If the profile matches the colony gestalt odor, entry is allowed; if not, aggression follows.

A chemical password like this seems secure because it is complex: not a single molecule, but a mixture of many compounds in particular ratios. Yet social parasites repeatedly evolve to exploit it. Alain Lenoir and Patrizia D'Ettorre (2001) reviewed strategies that are almost maddeningly clever. Some parasites practice *chemical insignificance*: they carry very little odor when first entering a colony, reducing the cues that would trigger rejection. An ant with a faint scent can slip past the threshold of attention, much as a person might pass a bouncer by blending into a crowd. Once inside, the parasite can acquire the colony odor by physical contact, grooming, or by synthesizing matching hydrocarbons. This is *chemical integration* or mimicry: the parasite gradually becomes chemically fluent in the colony's dialect.

The consequences can be stark. A parasite that passes as a nestmate may be fed, protected, even carried into safe brood chambers, where it can consume larvae or manipulate workers. The colony's own co-operative behavior becomes a resource to mine. Here, counterfeiting is not only about obtaining food; it is about redirecting labor. The parasite does not need to win a fight if it can reprogram the colony's algorithms of care.

Human analogies come easily, but the biological details are more interesting than the metaphor. Cuticular hydrocarbons are not arbitrary symbols; they are also functional components of the insect's waterproofing and physiology. This dual role can constrain how easily a colony can change its chemical profile to thwart parasites. A lock that is also part of the doorframe cannot be redesigned freely. Parasites exploit these constraints: they attack the edges of recognition, where the host cannot tighten security without paying a price.

Mutualisms between plants and insects offer a different kind of counterfeiting: not slipping into a guarded nest, but slipping into a contract. The yucca and the yucca moth are a famous example of an unusually intimate mutualism. The moth deliberately collects pollen and actively pollinates the yucca flower while laying eggs. The developing larvae then feed on some of the seeds. In effect, the moth provides a service (pollination) and receives a resource (a portion of the offspring) as payment. The plant can tolerate some seed loss because without the moth there would be few or no seeds at all.

But even in this tight partnership, free riding evolves. A. J. Tyre and J. F. Addicott (1993) documented facultative "cheating" behavior in which some moths lay eggs while transferring little or no pollen. Later, Olle Pellmyr and James Leebens-Mack (1996) described non-pollinating yucca moths that lay eggs but do not pollinate, exploiting the plant's reproductive machinery without paying the pollination fee. The important point is not that moths are morally flawed; it is that the opportunity exists because the plant cannot perfectly link egg-laying to pollination in real time. The moth can separate "take" from "give."

In such systems, chemistry often supplies the early steps: cues that guide oviposition, signals that indicate flower stage, odors that attract or repel. A cheater does not always need to mimic a molecule precisely; sometimes it only needs to arrive at the right time, or to avoid triggering a deterrent. If a plant uses a chemical cue to indicate "I am ready for a pollinator," any organism that can detect that cue can attempt to cash in, whether or not it will be a good partner.

This is where the energetic difference between honest signaling and theft becomes more than a slogan. Honest participation requires machinery that does work: enzymes, transport proteins, muscles, developmental programs. Deceptive participation often requires only *access* to the host's response. If a plant opens a nectar spigot when a specific odor is present, then producing the odor is cheaper than producing nectar. If a host feeds those who display a colony scent, then synthesizing or acquiring that scent can be cheaper than evolving into a cooperative worker. In economic terms, the cheater exploits a separation between the cost of *producing the signal* and the cost of *providing the service*. Evolution tends to narrow that separation when it becomes too dangerous, but it rarely eliminates it.

The most unsettling aspect of chemical counterfeiting is how it turns a host's best features into vulnerabilities. Sophisticated recognition, fine-tuned sensitivity, and rapid response are advantages in a world where partners are valuable and time is limited. Those same features make deception more profitable. A highly cooperative system often has strong pipelines for moving resources quickly once recognition occurs. In other words, cooperation builds high-throughput infrastructure. Counterfeiters aim for that infrastructure, not for the host's destruction. A parasite that kills its host too quickly burns down the bank it is robbing.

This is why deception so often leads to an arms race of information. Hosts change the molecular details of recognition, tighten thresholds, add layers of confirmation, or demand more costly proof. Counterfeiters respond by improving mimicry, learning new cues, or exploiting the host's own constraints. The result can be an odd mixture of stability and tension. From a distance, the mutualism persists for millennia, and the same partners keep meeting. Up close, there is constant pressure on the exact chemical words they use to say "friend."

Sometimes the host's best defense is not to become more suspicious in general, but to become more *conditional*. If the cost of screening is high, the host can begin with a small investment and escalate only when returns appear. The first sip of trust is cheap; the larger flow requires evidence. This kind of logic is common in biology because it is cheap and flexible. It does not demand perfect identification at the start; it demands performance over time. A counterfeiter can still slip in, but the payoff shrinks if the host can reduce the initial gift and increase it only when it is repaid.

Yet even conditionality can be hacked if the early evidence is itself chemical and therefore forgeable. A parasite can secrete compounds that simulate successful interaction, much as a fake receipt can satisfy a superficial audit. The deeper the partnership, the more complex these games can become, because the signals are woven into metabolism and development. In many mutualisms, the signals are not mere labels but triggers that reshape bodies: roots build nodules, insects change behavior, immune systems stand down. Once those switches exist, flipping them becomes a powerful act.

This is the core lesson of counterfeiting in nature: cooperation creates meaning, and meaning can be stolen. Molecules become words; words become promises; promises become pipelines for food, protection, and reproduction. The cheater's genius lies in recognizing that the host is not guarding a resource directly. The host is guarding a decision rule, encoded in receptors and neural circuits, that tells it when to give. Find the rule, imitate the input, and the output follows.

The uncomfortable elegance of it all is that no deception is permanent. Chemical mimicry succeeds only while it remains plausible, only while the host's recognition machinery continues to accept the forged key. And so the crowd keeps shifting: honest partners refining their cues, opportunists refining their counterfeits, both bounded by the costs of chemistry and the constraints of bodies. A mutualism is never just two species exchanging resources; it is also a living protocol, a language under pressure, spoken in compounds that can nourish, mislead, and, on occasion, betray.

4.2 Crime and Punishment

A forged signal can open the door once. The harder trick is to keep the door open, season after season, when the host is not a passive backdrop but a living accountant. Mutualisms persist because hosts are not simply generous; they are selective. They invest where returns are likely and withdraw where returns dry up. In human terms we reach for the language of morality—crime, punishment, cheating—because it captures the feeling of violation. Biologically, the same pattern can be described without indignation: a host that pays indiscriminately will be outcompeted by a host that pays conditionally. "Punishment" is often just resource management under uncertainty, sharpened by natural selection into something that looks uncannily like enforcement.

The main difficulty is practical. A partner's service is often invisible at the moment payment is made. A plant cannot watch a bacterium split nitrogen molecules in real time. A coral polyp cannot peer inside every algal cell to see how much sugar is being exported. A cleaner fish cannot perfectly monitor whether its cleaner is eating parasites or taking forbidden bites of mucus. The world is full of delayed payoffs and hidden actions, which is exactly the landscape where cheating pays. Sanctions evolve as ways to close that accounting gap, making the benefits of deception temporary and the costs of non-cooperation immediate.

A useful term here is *partner choice*. Even without intelligence in the human sense, hosts can bias their interactions toward better partners. Sometimes choice is literal: a client fish swims away from a dishonest cleaner. Sometimes it is physiological: a plant channels more carbon to one fungal strain than another. Sometimes it is brutal: a fig tree aborts a developing fig that has been overrun by non-pollinating wasps. The common theme is simple: cooperation is stabilized when hosts can (i) compare partners or detect poor performance and (ii) change the flow of resources accordingly.

The most iconic example is a relationship that begins underground, in darkness, but supports much of the world's agriculture: the legume–rhizobium mutualism. Legumes include peas, beans, clover, and soy. Their partners are soil bacteria called rhizobia. When the partnership works, rhizobia live inside root nodules and convert atmospheric nitrogen gas N_2 into ammonium, a form the plant can use to build proteins and DNA. In exchange, the plant supplies carbon compounds as food and provides a carefully controlled low-oxygen environment that allows the bacterial enzyme nitrogenase to function.

Nitrogen fixation is costly. Breaking the triple bond of N_2 is an energetic ordeal. If a rhizobium could live in a nodule, drink the plant's carbon, and *not* fix nitrogen, it would save energy and potentially reproduce faster. From the plant's point of view, that is an intolerable bargain: the plant has built a specialized organ, diverted sugars, and adjusted its own physiology to host a partner that is supposed to provide a scarce nutrient. If cheaters multiplied unchecked, nodules would become sugar sinks, and the mutualism would rot from the inside.

The surprise is that it often does not. Part of the reason is that legumes do not simply "invite" rhizobia and then hope for the best. They police.

E. Toby Kiers and Robert Rousseau (2003) offered one of the clearest experimental demonstrations that legumes impose sanctions when rhizobia fail to deliver. They worked with soybean and manipulated the conditions so that rhizobia inside nodules were prevented from fixing nitrogen—a way of making otherwise cooperative bacteria behave like cheats. The host plants responded by reducing the reproductive success of those bacteria, roughly halving it. Crucially, the mechanism appeared to involve oxygen: the plant restricted oxygen supply to the nodule, and the bacteria suffered.

To understand why oxygen is such an effective lever, it helps to appreciate the strange engineering of a nodule. Nitrogenase is highly sensitive to oxygen; too much oxygen inactivates it. Yet rhizobia still need oxygen for respiration to generate the ATP required for nitrogen fixation. The nodule therefore maintains oxygen at a low but usable level, a controlled trickle. This is managed by plant tissues that regulate diffusion and by oxygen-binding proteins such as leghemoglobin, which gives many nodules their pinkish color. The host has built a room where oxygen is kept on a dial, not a switch.

That dial becomes a weapon. If the plant senses that the nodule is not providing nitrogen, it can tighten the oxygen supply. The bacterial metabolism slows. Growth and reproduction falter. The nodule becomes, from the bacteria's perspective, a bad neighborhood: plenty of carbon may still arrive, but without enough oxygen it cannot be turned into energy, and the promised refuge turns into suffocation.

The word "senses" hides some elegant chemistry. Plants do not carry tiny nitrogen meters into each nodule; instead they read a set of internal cues. A well-functioning nodule exports nitrogenous compounds into plant tissues. If export drops, local concentrations shift. The plant's own nitrogen status changes, which influences hormone signaling and gene expression. Within the nodule, the balance of metabolites and oxygen consumption changes in ways that can be detected. The exact chain of detection is complex and still an active area of research, but the logic is robust: the host tracks *outcome* rather than *intention*. It cannot know whether a rhizobium is lazy or simply stressed, but it can respond to the net flow of nitrogen.

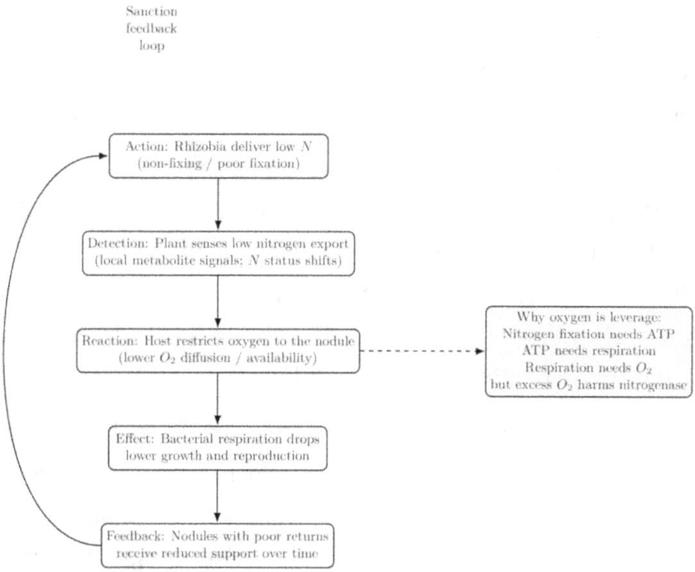

Sanctions like this are easy to admire because they look decisive, but they are not magic. A sanction must satisfy several constraints to shape evolution rather than merely harm everyone.

First, it must be targeted enough that cooperative partners do not pay too high a collateral cost. If cutting oxygen harms the plant as much as the bacteria, it is not a viable long-term strategy. The plant has to be able to withdraw support from one nodule without crippling its whole root system. The modularity of nodules matters. A nodule is a little compartment; it can be starved without starving the entire root.

Second, sanctions must affect the *reproductive success* of the would-be cheater. If the bacteria can reproduce freely before sanctions arrive, the damage may be too late. Timing matters. So does the biology of nodules, a point emphasized by R. Ford Denison (2000). In some legumes, rhizobia differentiate into bacteroids that may lose the ability to reproduce. In such cases, punishing them is like punishing sterile workers; it may protect the plant in the moment but does not necessarily select for more cooperative bacterial genotypes. Where bacteroids retain reproductive potential, sanctions can more directly change which strains leave descendants.

Third, the host needs a reasonably reliable cue. Not perfect, but good enough. A plant that punishes because a nodule is temporarily underperforming due to drought could end up harming its best partners. That risk pushes hosts toward sanctions that are graded and reversible rather than purely lethal. It also pushes mutualisms toward redundancy: multiple nodules, multiple partners, multiple chances.

There is a deeper insight here: what looks like "punishment" can be a form of prudent investment. Megan Frederickson (2013) argued that sanctions are sometimes framed too narrowly as retaliation against cheaters, when they may also arise as host responses to normal variation in partner quality. In a soil full of diverse microbes, some rhizobia will be better at nitrogen fixation than others even without any strategic deception. A plant that shifts resources away from low-performing nodules is behaving like an investor reallocating funds from a low-return project to a high-return one. Cheaters, if they exist, are then caught in the net of a broader optimization strategy.

This is not just semantic. It changes how one imagines the evolutionary pressure. If sanctions only work when cheaters are common, they might rarely be favored. If the same physiological machinery improves fitness by responding to ordinary variability, it can evolve even when true cheats are rare. Enforcement, in that view, is an emergent property of doing sensible biology under uncertainty.

The logic of sanctions becomes even more intuitive when moved above ground, to systems where the "host" is not a plant but an animal that can walk away. Cleaner fish mutualisms are famous for being both charming and fraught. In many coral reef systems, cleaner wrasses set up stations where larger "client" fish come to have ectoparasites removed. The cleaners gain food; the clients gain hygiene and reduced parasite load. But cleaner fish prefer to eat client mucus, which is more nutritious than parasites. That preference creates a built-in temptation: take a bite of mucus (cheat) rather than pluck a parasite (cooperate). The client feels the bite as a jolt.

Redouan Bshary and Alexandra Grutter (2005) explored how cooperation is maintained in this system by showing that clients can punish or switch partners. When a cleaner cheats, a client may chase it, disrupting feeding, or simply leave for another cleaning station. Either way, cheating becomes costly. The cleaner, in turn, learns to behave better under credible threat. The biology here is behavioral rather than biochemical, but the underlying currency is still chemistry: mucus and parasites are different foods, and the conflict is about who gets to eat which molecules.

The reef example matters because it shows that sanctions can operate on short time scales, in minutes rather than weeks, and that the principle of enforcement does not require sophisticated moral cognition. It requires only that a host can change the payoff to a partner based on recent outcomes. In a plant, that might mean rerouting oxygen. In a fish, it might mean a chase. Different machinery, same logic.

Plants, unable to chase, sometimes use a harsher tool: abortion. A fig is not just a fruit; it is also a nursery and a marketplace. Inside a developing fig, tiny fig wasps pollinate the internal flowers and lay eggs. Pollination benefits the tree; the wasp larvae benefit from developing seeds. But the system attracts exploiters: non-pollinating wasps that lay eggs without providing pollination, turning the fig into a nursery without paying rent. A tree that continues to invest in heavily exploited figs wastes resources that could have been used for more successful reproduction.

Recent work by Shang-Yang Lin and Bruno Di Giusto (2025) examined fig abscission—the tree dropping figs—as a defense against parasitism by non-pollinating fig wasps. Their experiments showed that lack of pollination and high parasitism increased fig abortion, and that heavily parasitized figs tended to be shed earlier. The effect is devastating to wasps developing inside: a dropped fig is often a death sentence. The tree's decision, if it is allowed that anthropomorphic word, is again based on outcomes: a fig that is not on track to yield sufficient reproductive value is cut off.

Abscission is an elegant kind of sanction because it is both punishment and triage. It punishes parasites by destroying their nursery. It also conserves the tree's investment, preventing continued allocation of sugars and water to a doomed structure. Once more, the line between retaliation and optimization blurs.

Across these examples, a pattern emerges in the chemistry of enforcement. Sanctions often target a bottleneck—a resource that is (i) essential to the partner's payoff, (ii) under host control, and (iii) cheap for the host to modulate. Oxygen in nodules is such a bottleneck. Continued fruit provisioning is such a bottleneck. Access to repeated cleaning opportunities is such a bottleneck. Hosts rarely need to invent an entirely new weapon; they use the valves they already possess.

This is also why sanctions tend to be local. Local control makes precision possible. A plant can restrict oxygen to a single nodule without suffocating its own tissues. A fig tree can drop a few figs without dropping its entire canopy. A client fish can punish one cleaner without abandoning cleaning entirely. Locality prevents enforcement from becoming self-harm.

Enforcement has an emotional resonance for humans because it feels like justice: wrongdoers suffer, cooperation survives. But nature does not guarantee a courtroom outcome. A sanction can fail. A cheat can spread if detection is noisy, if punishment is delayed, if hosts are constrained. Sometimes the host is trapped into continuing payment because the alternative is worse. Sometimes the partner can evolve countermeasures, such as producing enough of the right signal to avoid detection while still withholding service. In those cases, the host's enforcement machinery is forced into an escalating contest of information, trying to measure what matters rather than what is easily faked.

Even when sanctions work, they reshape the mutualism. A rhizobium lineage under strong host control may evolve to be more cooperative, but also more dependent on the host, because the host controls key resources and conditions. A cleaner fish under strict client punishment may become more attentive, but also more strategic, prioritizing clients who can punish and cheating those who cannot. Enforcement does not just eliminate conflict; it organizes it.

There is something oddly comforting in the idea that cooperation can be defended without kindness. A legume does not need to love its bacteria; it needs to allocate oxygen wisely. A fig tree does not need to hate a parasitic wasp; it needs to avoid wasting sugar. A client fish does not need to philosophize; it needs to leave when bitten. The chemistry of sanctions is not a moral story, yet it echoes the same lesson: agreements endure when breaking them carries a cost, and the cost can be imposed by something as simple, and as intimate, as closing a valve.

4.3 The Stability Paradox

A mutualism looks, at first glance, like an invitation to be exploited. If one partner is willing to hand over sugar, shelter, or transport in exchange for a service, why not take the payment and skimp on the service? The temptation is built into the logic. Yet the living world is not littered with the ruins of collapsed cooperation. Legumes still house bacteria. Figs still bargain with wasps. Corals, in calmer years, still glow with algal partners. The paradox is not that cheating exists, but that it so often fails to become the whole story.

Part of the answer is mundane: exploitation is harder than it appears. A cheater must get inside the partnership, avoid being detected, and reproduce fast enough to matter. That is already a tall order. A deeper answer is more interesting: many mutualisms are not fragile agreements held together by goodwill. They are dynamic systems with feedbacks, spatial structure, and shifting rules. They tolerate a certain amount of dishonesty the way a healthy body tolerates a certain amount of noise in its physiology. Collapse is only one of the possible outcomes, and not usually the most likely.

One reason is that mutualisms often behave like markets rather than marriages. The currency may be carbon and nitrogen rather than money, but the logic is familiar: rewards flow toward better offers. A plant that can choose among fungal partners, even imperfectly, does not need to eliminate every cheater to do well; it only needs to bias its investment so that good partners become more common around it. The cheater can exist, even persist, without taking over, because the host's decisions continually reshape the economic landscape in which the cheater is trying to thrive.

Benedicte Bachelot and Charlotte T. Lee (2018) modeled a form of this market behavior in arbuscular mycorrhizal symbioses, where plants trade carbon for mineral nutrients with fungi that colonize their roots. Their emphasis was on *dynamic preferential allocation*: the plant does not make a single once-and-for-all choice. It keeps adjusting. When nutrient returns from one fungal type are high, more carbon is routed that way; when returns are low, investment shifts elsewhere. This ongoing steering can maintain diversity of fungal partners and prevent any one strategy from monopolizing the system. Diversity is not just a side effect; it can be part of the stabilization mechanism, because it gives hosts options and forces symbionts to compete for payment.

This kind of feedback is especially powerful because it does not require a dramatic act of punishment. It is closer to turning down the volume than smashing the radio. A cheater that provides little benefit may still get a trickle of carbon, especially if the host cannot perfectly distinguish it from a mediocre mutualist, but it is starved of the large flows needed for dominance. In a fluctuating environment, where today's poor partner may be tomorrow's useful one, such gentle steering can also be safer than harsh policing.

There is also a second stabilizer, less intuitive but pervasive: the structure of space. Many textbook arguments about cheating assume a well-mixed world where everyone interacts with everyone else. Real mutualisms are rarely like that. Soil is a patchwork of microhabitats. A coral colony is a mosaic of polyps and light gradients.

Insect colonies have guarded boundaries and internal compartments. Spatial structure can slow the spread of cheaters, confine damage, and create safe neighborhoods where cooperators mostly meet cooperators.

Imagine a plant root as a city. It does not negotiate with "the soil microbiome" as a whole. It negotiates with the microbes that happen to be near a particular root hair on a particular day. If a cooperative strain establishes itself and helps the plant grow, the plant's root system expands, creating more habitat and more opportunities for similar partners. If a cheater establishes itself but provides little return, the local tissue may receive less investment, grow less, or be pruned. The consequences of a poor partnership are therefore partly local. A cheater cannot automatically convert the entire plant into a resource stream; it must repeatedly succeed at many points along the growing frontier.

Spatial structure does more than slow the cheater; it changes what selection favors. In a well-mixed scenario, a cheater can exploit cooperators across the whole population, taking benefits without paying costs. In a spatially structured scenario, cheaters are more likely to be surrounded by their own kind as they spread locally. That is a problem for them, because cheaters often depend on cooperators to maintain the system they exploit. A classic free-rider strategy can be self-limiting when it creates neighborhoods with too few honest partners to keep resources flowing.

Pattern formation can even create multiple stable states: patches where mutualists dominate and patches where they fail, coexisting in the same landscape. From the standpoint of everyday life, this is reassuring. It means cooperation can survive local disasters. A disease outbreak, a drought, or an invasion by cheaters can ruin some patches without necessarily tipping the entire system into collapse. The world becomes a quilt rather than a uniform fabric, and that patchiness is a form of resilience.

A third stabilizer is the coevolutionary treadmill sometimes nicknamed the Red Queen. The phrase comes from evolutionary biologist Leigh Van Valen (1973), who borrowed it from Lewis Carroll's character who must keep "running" to stay in the same place. In mutualisms threatened by counterfeiting, the "running" often involves chemical passwords. A host evolves receptors tuned to a partner's signals; a cheater evolves molecules that mimic them; the host evolves new thresholds, new blends, new confirmations. The overall mutualism can remain recognizable for long stretches of time even as the molecular details are constantly renegotiated.

This helps make sense of a fact that initially seems contradictory: some cheater lineages can persist for a long time without destroying the mutualism they exploit. The yucca moth system is an instructive case. Some yucca moths pollinate and lay eggs; others lay eggs without pollinating. Pellmyr and Leebens-Mack (1996) argued that non-pollinating yucca moth lineages have existed for long periods, not as brief anomalies. The mutualism persists anyway. One reason is that yucca plants have responses that reduce the cost of being exploited, such as aborting flowers with too many eggs or inadequate pollination, which limits cheater success. Another reason is that the ecological context matters: pollinating moths are still needed, and cheaters cannot necessarily replace them. The system can settle into a tense equilibrium in which exploitation exists but is contained.

It is tempting to picture stability as a fixed point: cheaters held at zero, cooperators triumphant. Many mutualisms are not like that. They can be dynamically stable, with fluctuations rather than stasis. When cheaters become common, hosts respond more strongly or shift partners, reducing cheater success; when cheaters become rare, the host's investment may become less stringent, allowing them to creep back. The system can oscillate, not because it is poorly regulated, but because regulation itself creates time-lagged feedbacks.

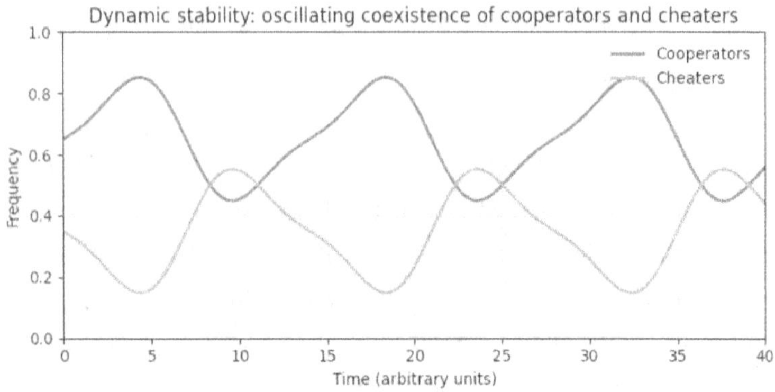

Oscillations are not just mathematical curiosities. They mirror something intuitive about enforcement: it is easier to detect and punish obvious exploitation than subtle underperformance. When cheaters are abundant and bold, they are conspicuous; hosts respond strongly; cheaters crash. When cheaters are rare, the pressure on hosts to maintain expensive surveillance weakens; tolerance increases; cheaters have room to return. Even a perfectly functioning sanction system can produce cycling if it carries costs and is adjusted over time.

A fourth stabilizer is that "cheater" is often a misleading label. Many partnerships run along a continuum of quality rather than a simple honest/dishonest split. A rhizobium strain may fix less nitrogen because it is genetically different, because the soil is phosphorus-poor, because oxygen conditions are unfavorable, or because the plant itself is stressed. A fungal partner may deliver little phosphorus in rich soil but be valuable for drought tolerance or pathogen resistance. When the environment changes, yesterday's free rider can become tomorrow's key ally. Under such conditions, hosts that respond flexibly to outcomes can do well without needing to classify partners into moral categories. Stability comes from adaptability.

This is where the earlier distinction between retaliation and optimization matters. If hosts are constantly reallocating resources based on returns, then even rare cheaters are kept in check as a side effect of ordinary decision-making. Frederickson (2013) emphasized that sanctions may not always be shaped mainly by a history of rampant cheating; they can emerge from selection on hosts to manage variable partner quality. Once those mechanisms exist, they also happen to make cheating less profitable. Stability, then, is not a special patch added onto an otherwise doomed mutualism; it is woven into the way organisms manage trade in an uncertain world.

Spatial structure and coevolutionary feedbacks also interact in a subtle way. Chemical recognition often evolves locally. A plant population in one region may be tuned to the microbial community it commonly meets; a parasite adapted to one chemical dialect may fail elsewhere. This creates a kind of geographic compartmentalization of the arms race, which again prevents a single super-cheater from sweeping through all populations. It also means that what looks like a stable mutualism at a global scale can be a patchwork of local skirmishes, each with its own temporary winners.

The emotional temptation is to ask for a final verdict: does cooperation win, or does cheating win? Living systems rarely provide that closure. The more accurate picture is of a working engine that includes friction as part of its operation. Some friction is harmful; too much can seize the machine. But some friction is also information: it reveals where the system is vulnerable and where it must be reinforced. Cheaters force hosts to become discriminating, and discrimination can make trade more efficient even among honest partners. Hosts force symbionts to compete, and competition can improve performance. A mutualism, in that sense, is not a truce that must never be violated. It is a set of feedback loops that keep the exchange worth doing, even when opportunists are always probing for the cheapest way to be paid.

If this sounds abstract, it has a concrete echo in everyday life: a garden bean plant pushing roots through soil does not encounter a single partner, but a shifting crowd. It cannot afford to be naive, but it cannot afford to be paranoid either. It does what successful organisms so often do. It samples, adjusts, compartmentalizes, and responds to outcomes, using chemistry as both language and ledger. The result is not perfect trust, but a practical stability that can survive noise, mistakes, and a certain amount of fraud without surrendering the benefits of cooperation.

Chapter 5

Underground Empires

Beneath the soil surface, a frantic chemical stock exchange operates continuously, trading carbon for minerals in transactions that define terrestrial life. We examine the molecular contracts signed between roots and microbes, revealing how plants actively recruit allies and manage their diverse portfolio of underground partners.

5.1 The Nitrogen Refinery

Air is mostly nitrogen. Every breath is taken in a sea of it. Yet for most living things, nitrogen in the form it occupies in the atmosphere is as useless as a gold bar to a shipwrecked sailor: abundant, present in every direction, and essentially unreachable.

The trouble is chemical stubbornness. Atmospheric nitrogen is N_2, two nitrogen atoms joined by a triple bond so strong that it behaves like a locked safe. Plants cannot crack it open directly. Animals cannot either; we rely on plants (or other animals) to do it for us. But proteins, DNA, chlorophyll, and many of the molecules that make life more than a bag of water are nitrogen-rich. A world full of nitrogen gas and hungry for nitrogen atoms has to solve an awkward problem: where does usable nitrogen come from?

For much of human history the answer was local and fragile. Farmers recycled nitrogen by returning manure to fields, rotating crops, or leaving land fallow. In the nineteenth century, European agriculture leaned heavily on imported guano scraped from Peruvian islands, an industry built on bird droppings, geopolitics, and a frantic awareness that soil fertility could be exhausted. The anxiety was real enough that when chemists and engineers finally learned how to make ammonia from N_2 and hydrogen at industrial scale, it felt like finding a hidden spring in a desert. Fritz Haber's laboratory breakthrough (1909) and Carl Bosch's engineering of high-pressure production soon after turned air into fertilizer and, indirectly, into bread. The modern world runs on that reaction.

Nature had been running a parallel reaction for hundreds of millions of years, but with a twist that still makes chemists jealous: it does not do it in steel reactors. It does it underground, in a living partnership that fuses diplomacy, architecture, and metabolic accounting into a single, quietly radical idea. Certain plants build a home for bacteria inside their roots. In exchange, the bacteria supply ammonia, a form of nitrogen that can be stitched into amino acids and nucleotides. It is not charity. It is not a casual friendship. It is a refinery contract, negotiated molecule by molecule, paid for in sugar.

Nitrogen fixation is expensive. Breaking N_2 is like splitting firewood with a teaspoon: possible only if you have a very special tool and the energy to swing it again and again. The bacterial enzyme that performs the job, nitrogenase, demands a steady stream of electrons and chemical energy, typically in the form of ATP. The cost is not paid in abstract units; it is paid in carbon compounds that the plant could have used to make leaves, seeds, or defenses. A nodule full of nitrogen-fixing bacteria is not a decorative accessory. It is a living organ that diverts a portion of the plant's photosynthetic income into an underground workforce.

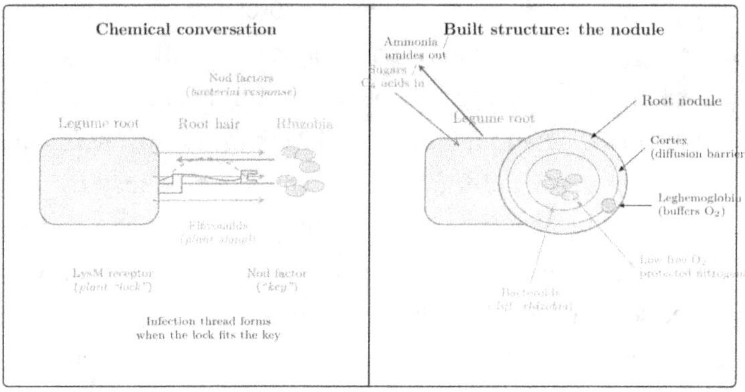

The plant does not merely tolerate this cost; it actively shapes it. That is one of the most counterintuitive features of symbiosis in roots: the plant behaves less like a passive host and more like an investor building infrastructure for a supplier, then monitoring performance. The nodule is a purpose-built organ, assembled only when the plant receives the right cues, and dismantled when the terms no longer make sense.

How does a root recognize the right partner in the chaos of soil, where countless microbes brush past every day? It begins with a chemical greeting. Legume roots release a suite of small molecules into the surrounding soil, including flavonoids and isoflavonoids. These are not random leaks. They are signals with a job: they recruit compatible bacteria and, crucially, instruct them to reveal themselves. When rhizobia detect the right flavonoid blend, a regulatory protein inside the bacterium switches on a set of genes called *nod* genes. The response is a secretion: Nod factors, which are lipochitooligosaccharides, small sugar-based molecules decorated with chemical side groups. Those decorations matter. They are the difference between being welcomed in and being treated as an intruder.

On the plant side, the root has receptor proteins embedded in the membrane of root hair cells. Many belong to a family called LysM receptor kinases, named for a protein domain that binds chitin-like molecules. That detail hints at the risk: chitin is also a component of fungal cell walls, and fragments of chitin can be read by plants as danger signals. Nod factors sit near that boundary between friend and foe. They resemble molecules that can trigger immunity, but their precise structure and the plant's matching receptor allow the message to be interpreted as an invitation rather than an alarm. It is a lock-and-key system, with the key cut to fit a particular lock. The plant is not asking, *"Is something there?"* It is asking, *"Are you the kind of partner that speaks this language?"*

The first visible consequence is a small contortion in a root hair, a curl that traps bacteria close to the surface like a hand closing around a rope. Inside that curl, the plant builds a narrow tube of cell wall material called an infection thread. It is not a wound; it is a guided corridor. Bacteria multiply within it as it extends inward, deeper into the root tissue, shepherded by the plant's own growth machinery. At the same time, cells in the root cortex begin dividing to form the nodule primordium, the early scaffold of the future organ. Infection and organ-building proceed together, coordinated by plant signals and bacterial replies, so that when bacteria arrive, a home is waiting.

The social logic is subtle. In ordinary pathology, microbes breach barriers and spread; the plant tries to seal, poison, and starve. In nodulation, the plant opens a controlled door and escorts the microbe through a controlled corridor. The plant is not surrendering; it is staging a managed invasion.

Once inside the developing nodule, rhizobia are not left to roam. They are engulfed by plant cells and enclosed in membrane-bound compartments called symbiosomes, where they differentiate into bacteroids. In many legumes, bacteroids become specialized, sometimes so altered that they cannot return to a free-living state. That can sound like domestication, and in a sense it is: a free-living bacterium becomes a worker in a plant-built refinery. But domestication cuts both ways. The plant supplies carbon and maintains the environment; the bacterium supplies fixed nitrogen. Both are constrained by the bargain.

The central machine is nitrogenase. It is one of biology's great chemical contraptions, a metalloenzyme assembled around iron and molybdenum (or other metals in some variants) that can reduce N_2 to ammonia. The reaction is exquisitely sensitive to oxygen. Oxygen damages the enzyme, yet the bacteria require oxygen to run their own respiration efficiently enough to generate ATP. This is the refinery's engineering headache: how do you run an energy-hungry, oxygen-sensitive process in a living tissue that, like all living tissues, needs oxygen?

The solution is an elegant compromise: keep free oxygen extremely low, while still allowing a controlled trickle sufficient for respiration. Nodules act as micro-oxic bioreactors. They construct diffusion barriers in their outer layers, slowing oxygen entry. They induce high-affinity bacterial respiratory enzymes that can scavenge oxygen at low concentrations. And, most famously, they produce leghemoglobin, a plant-made oxygen-binding protein that gives active nodules a pinkish tint. Leghemoglobin works a bit like hemoglobin in our blood: it binds oxygen tightly, buffering its concentration. Oxygen is not absent; it is channeled. Nitrogenase is protected not by starving the tissue, but by managing oxygen as a controlled reagent.

This engineering has a sensory feel if you have ever pulled up a healthy bean or clover plant and split a nodule open with a fingernail. A good nodule is often rosy inside. That color is not decoration. It is the signature of a working oxygen-management system, the difference between a refinery running at the right pressure and an idle building.

A plant pays for this refinery primarily with carbon compounds. Photosynthesis in the leaves produces sugars that are transported down to the roots. Within nodules, plants often deliver carbon to bacteroids as C_4-dicarboxylates such as malate or succinate, molecules that plug directly into bacterial energy metabolism. In return, ammonia produced by nitrogenase is rapidly assimilated by the plant, often converted into amino acids like glutamine or into transport forms such as ureides, depending on the plant species. Nitrogen moves upward from the root, becoming chlorophyll in leaves and protein in seeds, turning air into growth.

The partnership is ancient, but our understanding of it grew out of a nineteenth-century argument about where plants get their nitrogen. In the early 1800s, some scientists suspected plants could somehow use atmospheric nitrogen directly; others insisted nitrogen must come from the soil. The mystery sharpened around legumes, which seemed to enrich soils and thrive in poor ground. A decisive set of experiments by Hermann Hellriegel and Hermann Wilfarth in the 1880s showed that legumes grown in sterile, nitrogen-poor conditions failed unless they received a small amount of soil extract; when they did, nodules formed and the plants flourished. Something in the soil, alive and transferable, was enabling nitrogen acquisition. A few years later, Martinus Beijerinck isolated the bacteria responsible, strengthening the idea that microbes were not just decomposers but chemical specialists capable of feats that plants alone could not perform.

That historical arc matters because it reframes what counts as an organism's capability. The legume's ability to grow without added nitrogen is not simply a plant trait. It is a partnership trait, a property of two metabolisms fused into a single working unit. In agriculture, this fusion has been practical for centuries. Farmers did not need to know about Nod factors to plant clover as "green manure." They only needed to notice that fields behaved differently afterward. The chemistry was there all along, negotiated belowground.

Negotiation implies the possibility of cheating, and cheating is not hypothetical. In any market, a supplier might accept payment without delivering goods; a buyer might squeeze a supplier after investment has been made. Soil bacteria include strains that can invade nodules but fix little nitrogen. If legumes had no defenses against such freeloaders, the logic of evolution would be brutal: why pay for a refinery if unreliable workers can move in?

Plants respond with a mixture of partner choice and enforcement. The first line is selective entry: only bacteria that produce the right Nod factors tend to gain access, because the molecular lock is picky. Yet specificity alone cannot guarantee honesty. A bacterium can carry the keys and still underperform. So legumes also appear to monitor outcomes. Nodules that fix less nitrogen can receive less carbon, grow poorly, or senesce earlier. This resembles sanctions in human contracts: reduce resources to the partner that fails to deliver, making cheating less profitable. The mechanism can be indirect and physiological rather than moral. If the plant senses less fixed nitrogen, it may adjust oxygen supply or carbon flow in ways that disadvantage that nodule. The practical consequence is the same: over time, bacterial strains that fix nitrogen tend to be favored because they are rewarded with the resources needed to reproduce and persist.

The bacteria, for their part, are not inert employees. They live in a crowded soil economy where survival depends on flexibility. Some rhizobia can live freely, feeding on root exudates, and only shift into symbiosis when the right host appears. They carry gene circuits tuned to plant signals, and in many cases symbiosis genes can be moved between bacteria, spreading compatibility in a way that resembles the spread of a technology. From the bacterial perspective, the nodule is a protected niche rich in carbon but demanding in behavior: fix nitrogen, tolerate low oxygen, submit to differentiation, and accept close regulation by the host cell.

One of the most remarkable features of this arrangement is that it is not merely biochemical; it is developmental. The plant has evolved a program to build a new organ in response to microbial conversation. That is not a trivial adjustment. It means the plant has encoded in its genome the expectation of a partner, and it has linked that expectation to a physical architecture: infection threads, symbiosomes, oxygen diffusion barriers, vascular connections to export nitrogen, and regulatory circuits to decide how much to invest and when to stop.

If the refinery metaphor feels too industrial for a root, it helps to remember that industry borrowed its metaphors from biology as often as biology borrows from industry. Both are constrained by energy, materials, and risk. Both operate under trade-offs rather than ideals. The nodule is not a perfectly benevolent organ. It can be costly in drought, vulnerable to certain pathogens, and sensitive to soil chemistry. Too much available nitrogen in soil can suppress nodulation, because the plant has no reason to pay for what it can get cheaply. Too little phosphorus can limit how much energy the plant can generate and therefore how much nitrogen it can afford to fix. Even cooperation has an accounting department.

Yet when conditions align, the nitrogen refinery changes landscapes. Legumes can colonize young, poor soils and enrich them, seeding future plant communities with fertility. In managed fields, they can reduce the need for synthetic fertilizers, with consequences that ripple outward: fewer fossil fuels burned to produce ammonia, less nitrogen runoff into waterways, fewer algal blooms, less dead-zone expansion in coastal seas. A single biochemical alliance in a root can be felt, eventually, in the clarity of a lake and the cost of a loaf of bread.

It is tempting to imagine the story ends with ammonia delivered and the contract fulfilled. But the deeper point is stranger: the partnership turns air into tissue by turning conversation into construction. A few nanograms of flavonoids drifting into soil are answered by a few nanograms of Nod factors drifting back, and out of that exchange comes an organ that did not exist the day before. The plant's root becomes not just a foraging structure but a negotiating table, and the bacteria become not just residents of soil but licensed operators of a reaction that most life cannot perform. Underground, in darkness, chemistry becomes cooperation, and cooperation becomes food.

5.2 Login Credentials for the Wood Wide Web

A root is an explorer with blunt senses. It cannot see the pocket of phosphate a few centimeters away. It cannot smell, in any human sense, the mineral seam left behind by a decaying leaf. It grows, probes, and negotiates with whatever it touches. For most plants on Earth, that would be an intolerably inefficient way to live if they had to do it alone. Soil nutrients are patchy, often locked onto mineral surfaces, and sometimes present in forms that move through water too slowly for a root to catch. The solution that evolved, again and again, is not a better root but a borrowed body plan: threads of fungus that can slip between soil particles, squeeze into microscopic pores, and forage with a reach that makes even a dense root system look clumsy.

Mycorrhizae are these partnerships between plant roots and fungi. The term means "fungus-root," and the relationship is so widespread that it is easier to name plants that *do not* form mycorrhizae than those that do. In the bargain, fungi deliver mineral nutrients and water; plants pay in carbon. It is the same general logic as the nitrogen refinery—trade across a biological border—yet the details feel different because the partner is different. A bacterium lives as a single cell. A mycorrhizal fungus lives as a filamentous network, an organism that builds itself as it searches. When a fungus works with a plant, it does not simply move in; it connects.

The popular phrase "wood wide web" captures something real and something aspirational. Something real: fungal hyphae (thin filaments) can form extensive networks in soil, and some of those networks can connect multiple plants. Something aspirational: the idea that forests behave like intentional communities, sharing resources with moral purpose, sending messages along hyphal "wires" the way computers send emails. The truth is both less sentimental and, in many ways, more interesting. The chemistry of mycorrhizae is not a children's story about cooperation; it is an ancient authentication system that allows intimacy without catastrophe. Before any nutrient trade can happen, a plant must persuade a fungus that it is a suitable host, and a fungus must persuade a plant that it is not an invading pathogen. That persuasion happens in small molecules exchanged in the dark.

Start with the plant's problem. Phosphorus, one of the main reasons plants recruit fungi, is essential for DNA, membranes, and ATP, the molecule that cells use like spendable energy. Yet phosphate, the form plants take up, clings to soil minerals and can become nearly immobile. A root may sit beside plenty of total phosphorus and still experience starvation because the phosphate is chemically locked away. Many plants respond by changing their exudates-what they leak and secrete into soil-and by changing their architecture. But a more powerful response is to call for help.

The call is a family of compounds called strigolactones. They are plant hormones inside the plant, regulating branching and development, but outside the root they become signals. When a plant is short on phosphate, it tends to release more strigolactones into the surrounding soil. These molecules are active at very low concentrations and do not persist long; they are meant to be detected, not to accumulate. Their message is not a nutrient offer in itself. It is closer to a knock on a door: *a host is here, ready to negotiate.*

On the fungal side, strigolactones wake up a particular kind of fungus, the arbuscular mycorrhizal fungi (often shortened to AM fungi). These fungi are ancient, likely among the earliest partners of land plants, and their basic lifestyle is distinctive. They form large spores that can sit dormant, waiting. When strigolactones arrive, the fungus responds with renewed growth: spores germinate, hyphae branch more, and the chance of an encounter with a root rises dramatically. In a crowded soil, that is already a kind of sorting. Plants under nutrient stress are broadcasting a stronger invitation; fungi that are ready to cooperate are becoming more exploratory.

Then the fungus replies. It produces signaling molecules often called "Myc factors," a loose name rather than a single chemical. Some are lipochitooligosaccharides-molecules with a chitin-like backbone and lipid decorations-remarkably reminiscent of the Nod factors used by rhizobia. This resemblance is not cosmetic. It hints that plants and their microbial partners have been negotiating with related chemical alphabets for a long time. The plant recognizes these fungal signals with receptors on its root cells, some of them again belonging to the LysM family of receptors that can bind chitin-like compounds. Recognition triggers a conserved internal signaling route in the plant, involving characteristic calcium oscillations in the nucleus: an intracellular "handshake" that says, *stand down, accommodate this visitor.* It is a remarkable act of restraint. Plants have immune systems; their default response to microbial molecules is suspicion. Mycorrhizal authentication is an evolved exception—an immune truce with terms.

That truce is not a general amnesty. Many fungi are pathogens. Many soil microbes try to breach roots without giving anything back. So the plant's receptors act less like a motion sensor and more like a bouncer at a club: entry depends on the right credentials. The fungus must present a pattern of molecules that fits the plant's recognition machinery, and the plant must be in a physiological state that permits the partnership. A plant well supplied with phosphorus may be less welcoming; a plant under stress may be more permissive but also more vigilant about costs. Cooperation is not a mood. It is regulated.

Once authenticated, the fungus does something bold: it enters the root. For AM fungi, the key structure is the arbuscule, a branched, tree-like form built inside root cortical cells. It is not a simple invasion that ruptures the cell. The plant actively remodels its own membrane to wrap around the fungal branches, creating a large interface where exchange can occur. If this sounds delicate, it is. The arbuscule is short-lived, lasting days to weeks, after which it collapses and is digested. The plant repeatedly builds and dismantles these interfaces, as if installing temporary ports for trade.

The interface is where the "wood wide web" gets its plausible foundation: hyphae extend out of the root and into soil, increasing the effective absorptive surface area far beyond what roots could achieve alone. A hypha is much thinner than a root hair and can enter soil pores that roots cannot. It can bridge gaps of dry soil. It can forage in multiple directions while the root continues growing elsewhere. In a practical sense, it is like adding a second circulatory system dedicated to acquisition.

Phosphate is the headline nutrient. Fungal hyphae can absorb phosphate from soil and transport it back to the plant. Nitrogen can also be transferred in some contexts, and water uptake can be improved, especially in dry conditions. The currency plants pay is carbon: sugars and lipids derived from photosynthesis. This payment is not symbolic. Mycorrhizal fungi are hungry. Their external networks are living tissue that requires energy to build and maintain. A plant that feeds a fungus is diverting a meaningful portion of its photosynthetic budget belowground.

That budget matters because it defines the partnership's tensions. Mycorrhizae are often described as mutualisms, and in many circumstances they are. But the benefits are not guaranteed. If soil phosphorus is abundant, a fungal partner may provide little marginal help while still demanding carbon. If light is scarce, the plant may be carbon-limited and less able to afford fungal support, even if nutrients are scarce. Fungal species differ in how much they deliver and how aggressively they demand. The relationship can slide along a continuum between helpful and less helpful depending on conditions, like any economic deal subject to supply and demand.

This is where the "login credentials" metaphor earns its keep. Logging in is not the same as being in a utopia. It is permission to connect, not a promise of friendship. A plant that authenticates a fungus is granting access to its internal space; in return it expects services. The plant can modulate that access. It can control the lifespan of arbuscules. It can adjust the flow of carbon. It can alter the expression of transport proteins that determine how much phosphate it receives across the interface. Fungi, too, can adjust how much they transfer. Experiments have shown that fungal partners can retain nutrients rather than releasing them when carbon supply is low, accumulating phosphorus in their own structures like a warehouse holding inventory when the buyer is unreliable. In other words, the negotiation continues after the handshake.

The cultural popularity of the "wood wide web" owes a lot to imagery: the forest as a networked intelligence. That imagery has a history. In the late twentieth century, ecologists increasingly emphasized connections—food webs, nutrient cycles, feedback loops. Mycorrhizal networks seemed to offer a physical embodiment of interdependence: hyphae linking roots, carbon and nutrients moving among plants, potentially even chemical warning signals propagating through shared fungi. A forest, in this view, becomes less a collection of individual trees and more a communal organism.

There is genuine science behind parts of this picture. Shared fungal networks do exist. Carbon can move from one plant to another through pathways that involve fungi, and isotopic tracer studies have documented transfer under certain conditions. Plants can also influence each other indirectly through their effects on shared microbial communities. But the strongest popular claims—preferential nurturing of offspring by "mother trees," or a ubiquitous fungal internet deliberately distributing resources for the good of the forest—have been challenged sharply. A major perspective analysis by Justine Karst and Melanie Jones (2023) argued that several widely repeated claims about common mycorrhizal networks in forests were insufficiently supported or unsupported in the peer-reviewed literature, and that positive citation bias helped inflate confidence. A countervailing view has emphasized that evidence for belowground carbon transfer is real and accumulating, even if its ecological significance and its pathways are not always clear. The disagreement is not about whether fungi matter; it is about what kind of story the evidence justifies.

The chemistry helps keep the discussion grounded. A fungal network is not a fiber-optic cable installed for communication. It is living tissue built for fungal survival. Its primary obligation, in evolutionary terms, is to the fungus, not to the forest. If carbon moves between plants, it may be because the fungus is balancing its own internal economy, moving resources toward parts of the network that keep it fed and growing. If defensive signals appear to correlate between plants, it may be because both plants respond to similar cues, or because microbes alter the local environment in ways that affect both, rather than because a fungus is altruistically broadcasting a warning. These are not cynical interpretations; they are simply interpretations that take seriously the incentives and constraints of all partners.

A concrete example can make the abstractions feel less airy. Consider orchids. Many orchids, especially as seedlings, cannot photosynthesize enough to support themselves. Their dust-like seeds contain almost no stored food. Germination depends on obtaining carbon from fungi in the soil. In these cases, the relationship is not a simple trade of plant carbon for fungal minerals. It can be closer to parasitism from the plant's side: the plant is taking carbon, sometimes indirectly derived from other plants that feed the fungus. This phenomenon, called mycoheterotrophy, shows that fungal networks can indeed serve as conduits for carbon between plants, but not necessarily in a benevolent direction. The orchid is not being "helped" because the forest loves its diversity; it is exploiting an existing fungal economy, using the same molecular credentials that ordinarily permit mutualism to gain access to carbon when it cannot yet pay for itself.

Or consider a more familiar garden case: peas or beans planted in a nutrient-poor bed. They may form both rhizobial nodules and mycorrhizae, drawing nitrogen from bacteria and phosphorus from fungi. Gardeners sometimes notice that such plants can establish better than expected in poor soils, and that a subsequent crop does better after legumes have grown. The improvement is not magic. It is the compounded effect of multiple underground contracts: one that refines nitrogen from air, another that scavenges phosphorus through fungal reach. But even here the contracts can fail. Excess fertilizer can suppress nodules and reduce mycorrhizal colonization; pesticides and soil disturbance can disrupt fungal communities; drought can alter the cost-benefit balance. Underground cooperation is robust enough to have persisted across deep time, yet sensitive enough to be broken by a few blunt human interventions.

The authentication story also explains an odd-seeming danger: the same chemical signals that recruit helpful fungi can be hijacked. Strigolactones, for example, do not only attract mycorrhizal fungi. They can stimulate germination of parasitic plants such as *Striga* (witchweed), which detect strigolactones as a cue that a host root is nearby. In parts of sub-Saharan Africa, *Striga* devastates crops like sorghum and maize, essentially eavesdropping on a plant's attempt to log in to a beneficial network and using the message as an attack vector. The incident is a reminder that chemical "passwords" are not secret in the way digital passwords are; they are broadcast into a public environment. Any organism that evolves the right receptor can listen.

If the fungal network is a kind of infrastructure, then credentials are not a one-time key but a continuously verified identity. The plant must keep signaling its readiness to host; the fungus must keep presenting the molecular cues that prevent immune escalation. At the physical interface, the plant wraps the fungus in a membrane that resembles a diplomatic cordon: close contact, intense exchange, strict supervision. Transport proteins embedded in that membrane determine how phosphate and other ions move into the plant, and how carbon-rich compounds move to the fungus. Even without knowing every molecular detail, the logic is clear. The plant and fungus have built a controlled border crossing.

That control is why mycorrhizae can be so widespread. A root that let fungi enter indiscriminately would be destroyed. A fungus that could not enter at all would starve in a world where plants hold the richest carbon supplies. The compromise is conditional intimacy:

connect, but only with credentials; exchange, but only at monitored interfaces; grow, but accept inspection.

Forests, then, are threaded with potential connections, some realized and some not, always contingent. The most responsible way to think about the "wood wide web" is not as a single benevolent network but as a dynamic mesh of partnerships, overlapping and dissolving, shaped by soil chemistry, season, species identity, and simple economics. A fungal filament does not need to carry an ethical message to matter. It only needs to move a phosphate ion from an inaccessible surface to a hungry root cell, and the result can be a tree that grows a little faster, casts a little more shade, drops a little more leaf litter, and changes the lives of everything beneath it.

In the end, the most astonishing feature of these networks is not that they make plants cooperative saints. It is that they make plants capable. By outsourcing reach to fungal threads, a stationary organism extends itself into the soil's fine structure, turning mineral scarcity into a solvable logistical problem. The credential exchange—strigolactones out, fungal signals back—looks modest on paper, just a few molecules traded in vanishing concentrations. But it grants access to an underground workforce that can be older than the plant itself, already woven through the soil, waiting for the right chemical knock to wake, branch, and connect.

5.3 The Wisdom of the Mob

Soil looks still until you remember what scale you are missing. Between a grain of sand and a splinter of decaying leaf there are water films, air pockets, and chemical gradients so steep they would count as cliffs in a microbial world. Bacteria live on those cliffs. They do not drift in a well-mixed soup; they crowd into crevices, cling to roots, and build slimy cities on surfaces. If roots are negotiating tables and fungi are the long-distance carriers, bacteria are the street-level politics: quick alliances, sudden betrayals, and a constant struggle to decide when it is worth acting together.

Acting together is the trick. A single bacterium can secrete an enzyme, but if the enzyme diffuses away it may help competitors more than itself. A single bacterium can try to attack a root, but plants detect damage quickly and respond with immune chemistry. A single bacterium can make a biofilm, a protective matrix of sugars and proteins, but it cannot build a wall by itself. Many microbial strategies pay off only if lots of cells commit at once. The problem is coordination without a leader.

Bacteria solve this with quorum sensing, a chemical census that lets them estimate how many neighbors are nearby and adjust their behavior accordingly. The basic idea is simple enough to explain at a dinner party: bacteria release small signaling molecules into their environment. As the population grows, the molecules accumulate. When the concentration crosses a threshold, bacteria interpret that as "we are many," and they switch on a new set of genes. The details are exquisitely diverse-different molecules, different receptors, different circuits-but the logic is recognizably democratic. No one bacterium decides. The group decides by building up a shared chemical vote.

The most famous quorum-sensing story began not in soil but in the sea, in a glowing bacterium associated with the Hawaiian bobtail squid. In the 1970s, J. Woodland Hastings and colleagues helped establish that *Vibrio fischeri* (now often called *Aliivibrio fischeri*) produces light only when it reaches high density inside the squid's light organ. Individual bacteria floating in seawater stay dark. Pack them together, and they glow. The discovery was a gift to biology because light is an unmistakable readout: the collective switch is visible. The chemical that mediates the switch is an acyl-homoserine lactone (AHL), one of a broad family of quorum-sensing molecules used by many Gram-negative bacteria.

Soil bacteria also use AHLs, along with other signals such as small peptides (common in Gram-positive bacteria) and a molecule called AI-2 that some researchers have framed as a potential cross-species "lingua franca." The point is not that all bacteria share one language-they do not-but that bacteria have repeatedly evolved ways to convert local density into a chemical measurement. That measurement is then linked to behaviors that are expensive, risky, or pointless unless performed en masse.

Biofilm formation is a classic example. A biofilm is a community of microbes embedded in a self-produced matrix, often rich in polysaccharides. In soil, biofilms can coat root surfaces or cling to organic particles. They trap water, resist being washed away, and protect bacteria from stressors such as drying or toxic compounds. But biofilms are costly: the matrix is built from carbon and energy that could have been used for growth. If only a few bacteria invest while others freeload, the investors lose. Quorum sensing provides a solution: do not start building until enough partners are present that the wall will actually stand, and freeloaders are diluted by the collective output.

Virulence can be quorum-controlled for similar reasons. For a pathogen, launching an attack too early can be self-defeating. A plant can detect a small invasion and localize defenses. A synchronized attack by many cells, producing enzymes that break down cell walls or toxins that suppress immunity, can overwhelm a host before it mounts an effective response. This does not make bacteria villains in a moral sense; it makes them strategic. Soil is a competitive arena. Timing matters.

The same chemistry can also underwrite protection rather than attack. Many bacteria that colonize roots help plants by suppressing pathogens, producing antibiotics, competing for space, or priming plant immunity. These beneficial behaviors can also be density-dependent: antibiotics are costly, and if produced at low density they may be wasted. A bacterial community that "waits" until it is sufficiently established before deploying its chemical defenses is behaving like a neighborhood watch that does not ring the alarm for every passing stranger, but does mobilize when the crowd becomes threatening.

Plants, meanwhile, are not passive hosts of this microbial politics. A root is constantly shedding compounds into soil: sugars, amino acids, organic acids, phenolics, and many other molecules. Some are food. Some are signals. Some are weapons. Root exudates can selectively favor certain microbes, not by conscious choice, but by creating a chemical environment where particular metabolisms thrive. A plant under attack can alter its exudation profile, changing who gathers around its roots. This phenomenon has been called the "cry for help" model: the plant, stressed by a pathogen or herbivore, leaks or secretes compounds that recruit microbes capable of fighting back.

There is evidence for such recruitment in several plant systems, including cases where disease suppressiveness can persist in soil after an outbreak, a kind of microbial memory. Gardeners and farmers sometimes notice it empirically: a patch of soil becomes "sick" for a particular crop, or, conversely, becomes resistant after a disease episode. The reasons can be many-nutrient depletion, pathogen build-up, changes in pH-but microbial community shifts are often part of the story. A pathogen can restructure the microbial neighborhood, and the neighborhood can restructure the pathogen's prospects.

Quorum sensing connects to this because it provides a lever for both plants and microbes. If bacterial cooperation depends on signal molecules, then interfering with those molecules can change group behavior without killing anyone outright. And plants appear to do exactly that. Some plants secrete compounds that mimic bacterial quorum signals, confusing the bacterial census. Others produce enzymes or reactive chemistry that degrade signals, effectively lowering the perceived density and delaying collective action. This is a sophisticated form of defense because it targets coordination rather than survival. Killing bacteria outright selects strongly for resistance and can harm beneficial microbes. Disrupting a quorum can disarm harmful strategies with less collateral damage, at least in principle.

The idea of disarming rather than killing has influenced human medicine as well. Researchers have explored quorum-sensing inhibitors as anti-virulence therapies, aiming to make pathogens behave as if they are alone. Soil is where many of the relevant microbes evolved, and it is where many antibiotics and signaling inhibitors were first discovered: a reminder that pharmaceutical inspiration often begins as ecological observation.

Yet the "wisdom of the mob" is not always wise. Quorum sensing can amplify destructive behavior just as easily as helpful behavior. It can produce blooms of slime that clog plant tissues, coordinated secretion of toxins, or collective migration that overwhelms a root's defenses. The key feature is not virtue but coordination: a population acting as a unit once it has enough members to matter.

Coordination is only half the story. The other half is historical contingency: who gets to the root first, and what they do once they arrive. Ecologists call this priority effects. In soil, the first colonizers of a root can shape the conditions for everyone who comes later. A benign bacterium that quickly forms a biofilm can physically block pathogens from attaching. A pathogen that establishes early can damage the root and change exudation patterns, recruiting allies or creating niches that favor opportunists. Even when two soils contain many of the same microbial species, the order of assembly can push their communities toward different stable states, like two cities with the same population but different laws because of a different founding history.

This matters for any attempt to engineer plant microbiomes. It is tempting to imagine that adding a "beneficial" bacterial strain to soil will reliably improve plant health. Sometimes it does. Sometimes the strain fails to establish, outcompeted or simply arriving too late. The root is not an empty apartment waiting for a tenant; it is a crowded market with contracts already signed. Priority effects can make microbiome interventions unpredictable unless they take timing seriously.

The soil microbiome also contains keystone taxa: microbes whose influence is disproportionate to their abundance. A keystone bacterium might produce a compound that shapes the local pH, freeing nutrients or inhibiting certain competitors. It might degrade a quorum-sensing signal, altering the behavior of many other bacteria. It might be a hub in metabolic cross-feeding, producing a vitamin that others require. Remove such a taxon and the community can reorganize dramatically, even if the total number of microbes stays similar.

At the same time, soil often exhibits functional redundancy: many different microbes can perform broad tasks such as decomposing simple sugars or respiring common compounds. This redundancy can buffer ecosystems against disturbances that reshuffle species. If one decomposer declines, another can fill the gap. But redundancy is not uniform. Narrow functions-specific steps in nitrogen cycling, for instance-may be carried out by fewer taxa. Those steps can become bottlenecks and points of fragility. A soil can be robust in its overall metabolism and still vulnerable in a specific chemical conversion that plants depend on.

All of this unfolds at the scale of microns, yet it scales up to outcomes farmers and foresters care about: disease outbreaks, fertilizer efficiency, drought tolerance. The chemistry that bacteria use to coordinate-AHLs, peptides, and other autoinducers-can influence whether a pathogen becomes a minor nuisance or a crop failure. The timing of microbial assembly can influence whether a seedling establishes or withers. The presence or absence of certain microbial functions can influence whether nutrients stay available or leach away.

It also changes how one thinks about plant individuality. A plant is not merely a genome expressed in green tissue; it is a host managing a crowd. The management is partly chemical, partly physical, partly immune. Roots exude compounds that serve as food and as signals. They present receptors that interpret microbial molecules as friend or foe. They can alter their own physiology-more lateral roots, more hairs, different transporters-to respond to the microbial neighborhood. In return, microbes respond not only to the plant but to one another, voting themselves into cooperative states when their numbers justify it.

A concrete scene makes the abstractions feel tangible. Imagine a young wheat plant in a field after rain. The soil is moist, oxygen levels are patchy, and root exudates form a halo of sugars and amino acids. Bacteria swarm toward the feast. Early arrivals attach and begin producing sticky polymers, forming microcolonies on the root surface. As the microcolonies grow, quorum signals rise. At a certain density, genes switch on: a biofilm thickens, enzymes are secreted, perhaps antibiotics are produced that suppress a fungal pathogen trying to establish nearby. The plant, sensing subtle molecular patterns-fragments of microbial cell walls, changes in ion flux, perhaps even bacterial signals themselves-adjusts its exudates. It may feed allies more generously or deploy compounds that keep potential attackers disorganized. None of this requires intention. It requires evolved circuitry on both sides, tuned by the repeated reality that in soil, solitary action is often wasted.

There is an irony here. Quorum sensing is sometimes described as communication, as if bacteria are chatting. But it is closer to measurement. The signal molecules are not primarily messages about the outside world; they are a readout of the bacteria's own abundance. In that sense, quorum sensing is a form of collective self-awareness. It lets bacteria behave differently when they are a handful versus when they are a crowd.

And crowds change what is possible. A crowd can build physical structures. A crowd can overwhelm a defense. A crowd can maintain a stable chemical environment within a biofilm, with gradients of oxygen and nutrients that allow different metabolisms to coexist centimeters away from one another. A crowd can even act as a memory, persisting in soil and influencing the next plant that grows there, long after the original root has died.

This is the underground empire that is easiest to overlook because it is so small and so familiar. Bacteria are often treated as background noise: countless, interchangeable, essentially anonymous. Quorum sensing, priority effects, and microbial gatekeeping reveal the opposite. A soil microbial community is structured, responsive, and capable of coordinated behavior that emerges only when enough individuals commit to it.

In the end, the most striking feature of bacterial consensus is its humility. It does not promise harmony. It does not even promise stability. It offers a way for tiny organisms to avoid wasting their efforts, to time their risks, and to make the most of being many in a world where diffusion can erase the work of the few. A root growing through that world is never alone; it is constantly brushing against collective decisions made in molecules.

Chapter 6

Glass Houses in a Warming Sea

Coral reefs stand as the ocean's most spectacular paradox: massive stone cities constructed by soft-bodied animals in nutrient-poor deserts, powered entirely by borrowed solar energy. We will descend into the cellular machinery of this partnership to understand how a precise exchange of waste and sugar builds ecosystems, and why a mere rise of two degrees transforms this ancient engine of cooperation into a mechanism of self-destruction.

6.1 Solar-Powered Stones

A coral reef looks like geology that has somehow learned to breathe. The ridges and buttresses are stone, yet they grow, repair themselves after storms, and keep their colour only as long as living tissue remains draped over the skeleton. A reef is not a pile of rocks that happens to be inhabited; it is a construction project run by animals. The builder is the coral polyp, a small cnidarian that resembles a sea anemone living in a cup of its own making. The surprise is how such a modest creature can afford to lay down so much limestone in waters that are famously poor in nutrients.

Stand on a tropical beach and the sea can look impossibly clear, as if somebody has filtered it. That clarity is not just an aesthetic detail; it is chemistry with ecological consequences. Clear tropical water is often oligotrophic: low in dissolved nitrogen and phosphorus, the fertilisers that fuel plankton blooms. In many oceans the richest ecosystems sit under nutrient-rich upwelling, or near rivers that bring land-derived minerals. Reefs thrive in the opposite sort of place. They flourish where the surrounding water is, from a nutrient perspective, nearly barren.

The old puzzle of reef science was how such productivity could arise from such scarcity. A major part of the answer is that a coral is not working alone. Inside the coral's tissue live photosynthetic dinoflagellates from the family Symbiodiniaceae, popularly called zooxanthellae. They are tenants, but also engines: they harvest sunlight and convert it into chemical energy. The coral, in return, provides the algae with a protected home and an unusually generous supply of the raw materials photosynthesis needs. The reef is built by this partnership's internal economy, a tight loop of traded molecules.

A useful way to picture that economy is to think less about "feeding" and more about bookkeeping. The coral polyp eats like an animal: it can catch plankton with stinging cells, and it can absorb dissolved organic compounds. But in clear tropical water those meals are intermittent. Meanwhile, the algal symbionts behave like a miniature garden. When light is abundant, they manufacture carbohydrates from carbon dioxide and water. Those carbohydrates are not kept entirely for themselves. A substantial fraction of the photosynthate leaks or is actively transferred to the host, often as sugars and other small carbon-rich compounds. This translocated carbon pays the coral's bills: respiration, mucus production, reproduction, and, crucially, calcification.

There is a historical moment when this "leak" stopped being an intuition and became something measured. In the late 1960s and 1970s,

Leonard Muscatine and colleagues used radioactive carbon tracers to track the movement of photosynthetic products from symbionts into host tissues. By labelling inorganic carbon and then following where the label went, they showed that the algae were not simply co-existing passengers; they were financiers. In a classic synthesis with James Porter (1977), Muscatine framed reef corals as mutualisms adapted to nutrient-poor environments: conservation of scarce nutrients, internal recycling, and a striking increase in skeletal building in the light. That last observation, "light-enhanced calcification," remains one of the most vivid signatures of the partnership. Limestone growth, which seems at first like a purely inorganic act, accelerates when the coral's resident algae are photosynthesising.

To see why, it helps to zoom in from reef to polyp to cell, and finally to the thin layer where chemistry turns into architecture. A coral's skeleton is mostly calcium carbonate ($CaCO_3$), commonly deposited in the crystal form aragonite. The coral lays this down in a controlled space between its living tissue and the existing skeleton, sometimes called the calcifying fluid or subcalicoblastic space. The coral does not simply allow seawater to precipitate rock at random. It manipulates the local chemistry: concentrating calcium ions, adjusting acidity (pH), and controlling the availability of dissolved inorganic carbon. In effect, it creates a small, regulated micro-environment where precipitation becomes favourable.

Precipitation of $CaCO_3$ is a balancing act among ions and acid–base chemistry. In seawater, most inorganic carbon is not present as CO_2 but as bicarbonate (HCO_3^-) and carbonate (CO_3^{2-}), and the proportions depend strongly on pH. At higher pH, more carbon exists as carbonate, the form that combines with calcium to make $CaCO_3$. Corals can elevate the pH in their calcifying space relative to ambient seawater, shifting the chemistry toward carbonate and making skeleton formation easier. This is not magic; it is energy-dependent transport: moving protons, ions, and carbon across membranes, keeping the right gradients in the right places. Chemistry is doing the work, but biology is paying the energy cost.

That is where the algae matter. Photosynthesis consumes CO_2 and produces O_2 and carbohydrates. The act of removing CO_2 from a microenvironment tends to raise pH, because CO_2 in water is linked to carbonic acid and thus to acidity. When symbionts draw down CO_2 in the coral's tissues, they can influence local acid—base conditions and the distribution of inorganic carbon species. At the same time, the sugars delivered to the host are metabolic fuel. The coral can burn them to power ion pumps and transporters that shape the chemistry of the calcifying fluid. Light, then, is translated into limestone by two routes at once: direct chemical consequences of CO_2 uptake and indirect energetic support for active regulation.

The partnership's nitrogen economy is just as important, and in oligotrophic seas it may be the part that most resembles a closed-loop household. Nitrogen is often the limiting nutrient for algae: they need it to build proteins, nucleic acids, chlorophyll, and the machinery of photosynthesis itself. Yet tropical seawater can be remarkably low in dissolved inorganic nitrogen. Corals, as animals, have a steady trickle of nitrogenous waste from their own metabolism. A familiar form is ammonia (NH_3) or its protonated form ammonium (NH_4^+), depending on pH. To a solitary animal, nitrogenous waste is something to excrete. To an algal symbiont, it is fertiliser.

Corals retain and recycle this waste within their tissues, and the algae take it up to make amino acids and other nitrogen-rich biomolecules. This does two things at once. It supplies symbionts with a critical nutrient in a nutrient-poor environment, allowing them to keep photosynthesising at high rates. And it prevents the host from losing nitrogen to the surrounding water. The coral's tissue becomes a kind of internal compost heap, but a clean one: waste is captured, converted, and returned as useful biomass and, eventually, as more photosynthate.

The exchange is not a charitable donation from algae to coral. It is a bargain between two metabolisms that need each other's by-products. The coral's respiration produces CO_2, the algae's favourite carbon source. The coral's waste produces NH_4^+, an excellent nitrogen source. Meanwhile, photosynthesis produces O_2 that the coral can use for aerobic respiration, and sugars that can be burned or converted into lipids and other storage molecules. The coral can be imagined as a landlord that pays its tenants partly in utilities: carbon dioxide and nitrogen. The tenants pay rent in electricity and groceries: oxygen and sugars.

What makes the bargain so powerful is proximity. The algae are not floating nearby; they reside within the coral's tissues, often inside the host's cells in membrane-bound compartments. Diffusion distances are tiny. A molecule of CO_2 produced by the coral's mitochondria does not have to find its way through centimetres of seawater; it can be captured within micrometres by a symbiont chloroplast. Likewise, the oxygen produced by photosynthesis is immediately available to host cells. In biochemical terms, the partnership is an arrangement that minimises leakage. The coral is not fertilising the ocean; it is fertilising its own internal garden.

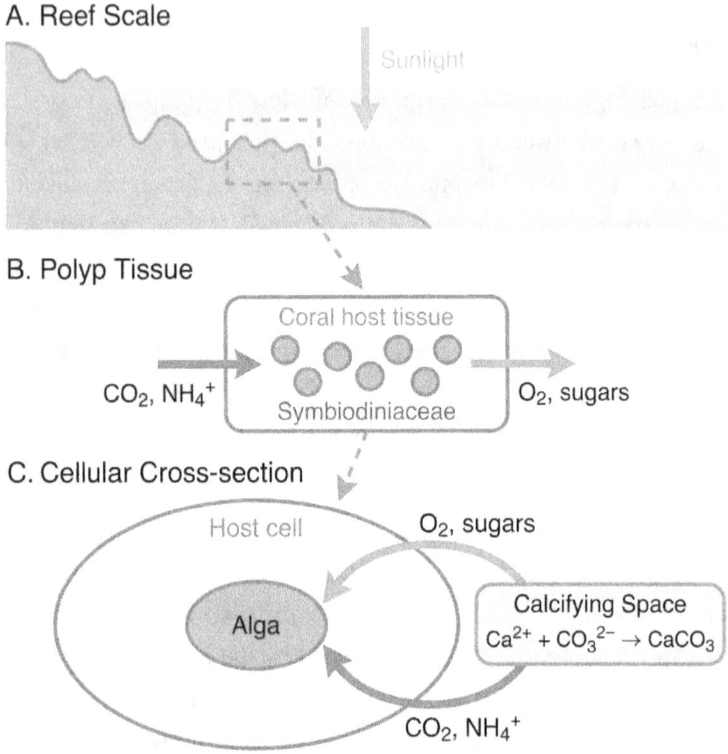

The diagram looks neat in two colours, but in a living coral the exchange is messier, negotiated moment by moment. The host controls many of the conditions the algae experience: light levels (by pigment concentration and tissue thickness), carbon supply, and nitrogen availability. The algae, in turn, respond to these constraints by changing how much carbon they keep versus how much they release. This makes the symbiosis feel less like a fixed machine and more like a managed micro-economy, with incentives and checks.

One of the simplest incentives is scarcity itself. If nitrogen is limited inside the coral tissue, algal symbionts struggle to build new cellular machinery. In that state, they may channel more of their photosynthetic carbon into compounds that are exported to the host, because their own growth is constrained. If nitrogen becomes plentiful, the algae can invest in division and biomass, potentially retaining more carbon for themselves. From the coral's perspective, this can be risky: too much nitrogen can turn cooperative tenants into self-investing ones. The coral cannot easily evict them one by one under normal conditions, but it can regulate the internal availability of nitrogen and carbon in ways that bias the relationship toward translocation rather than hoarding. Even in a partnership that is mutualistic overall, each partner has a lever it can pull.

The carbonate skeleton itself adds another layer to the bargain. It is tempting to treat calcification as a mere by-product of having extra energy, but it functions like infrastructure. A polyp embedded in a rigid skeleton can lift itself into better light and better flow. Better flow delivers more dissolved inorganic carbon and removes waste; better light powers more photosynthesis. The skeleton, then, is not just the house the coral lives in; it is scaffolding that improves the productivity of the very partnership that builds it. Over time, that feedback produces the familiar reef shape: ridges, plates, and branching thickets arranged to capture light and water movement, rather like a city skyline built to harvest sun and wind.

It also produces vulnerability in a strangely intimate way. Because the skeleton is deposited beneath living tissue, the border between animal and mineral is thin. A healthy coral polyp is often only a few millimetres thick; much of what we call "coral" is stone laid down by countless previous generations of tissue. The living layer is a skin of active chemistry stretched over architecture. When it thrives, it can cover bare skeleton, seal small wounds, and expand. When it struggles, the stone remains as evidence of what once was alive.

One concrete example of how tightly chemistry and ecology are tied comes from the daily rhythm of reefs. Measurements on many reefs show that the chemistry of the water overlying coral communities shifts between day and night. In daylight, photosynthesis draws down CO_2 and can raise local pH; at night, respiration dominates, releasing CO_2 and lowering pH. The reef breathes in a chemical sense, and the swing can be large enough to matter for calcification dynamics. For a coral polyp, this means the conditions for skeleton building are not constant: the microenvironment can favour precipitation during the day and ease off at night. The reef's "stone-making" is therefore, in part, a solar schedule.

The wonder of this arrangement is not that it breaks the laws of chemistry, but that it exploits them. In the open ocean, a molecule of nitrogen waste is diluted into an enormous volume and lost to the organism that produced it. A molecule of CO_2 exhaled by an animal is similarly dispersed. In corals, these losses are reduced by architecture: by living in colonies, by retaining symbionts within tissues, and by building skeletons that influence flow and light. The reef's abundance is built from not wasting what is scarce.

Seen from far enough away, a reef can look like a fixed object, a "formation." Yet its limestone is the hardened record of a metabolic conversation repeated billions of times: a polyp releasing CO_2, an alga using it; an animal excreting NH_4^+, an alga assimilating it; a symbiont producing sugars, a host turning them into ion gradients; calcium and carbonate meeting in a controlled pocket to become crystal. Sunlight enters the water as photons and leaves, partly, as stone.

6.2 The Toxic Divorce

A bleached coral can be beautiful in the way a hospital corridor is clean: bright, stripped, wrong. The white is not a new colour so much as an absence. The coral's limestone skeleton has always been pale; it is usually hidden under a living veneer tinted gold, brown, or green by millions of symbiotic algae. When those algae vanish, the stone shows through like bone under skin.

Bleaching is often described as corals “getting too hot,” as if an animal simply faints in warm water. That makes it sound passive and almost gentle. At the scale where bleaching begins, nothing is gentle. The coral is responding to an internal chemical emergency. The partnership that usually turns sunlight into sugars can, under stress, turn sunlight into poisons. Expelling the algae is closer to throwing a smoking battery out of a house than to abandoning a helpful roommate. It is a desperate act that may save the host cell in the moment while starving the animal over the weeks that follow.

The key to this reversal is that photosynthesis is not a single reaction but a chain of carefully timed transfers of energy and electrons. In calm conditions, light energy absorbed by pigments is channelled into a controlled flow that ends with carbon fixed into carbohydrates. The chain is robust, but not infinitely so. Heat makes molecules jiggle; membranes become more fluid; protein complexes that normally hold their shapes can slip out of alignment. Meanwhile, the sun does not dim just because the machinery is struggling. Excess light energy has to go somewhere.

When that energy cannot be used to build sugar, it spills into oxygen chemistry. Oxygen, which is normally the benign product of photosynthesis, becomes the raw material for reactive oxygen species, often abbreviated ROS. These include molecules such as superoxide and hydrogen peroxide: forms of oxygen that are chemically eager, prone to damaging lipids, proteins, and DNA. In ordinary cells, small amounts of ROS are inevitable and manageable. Cells carry antioxidant defences—enzymes and small molecules that neutralise these reactive forms before they cause harm. Under thermal stress in the bright tropics, symbionts can produce ROS faster than either partner can mop them up. What was a profitable exchange becomes a toxic leak.

The danger is amplified by the very intimacy that made the symbiosis productive. In open water, a burst of reactive molecules would be diluted. Inside coral tissue, the symbionts are packed close to host cells; diffusion distances are microscopic. Oxidative damage can accumulate locally, in the membranes that separate symbiont from host, and in the host's own cellular compartments. A stressor that begins as a glitch in algal photosynthesis becomes a broader cellular crisis, triggering host signalling pathways that resemble inflammation and immune-like response.

One of the unsettling lessons from bleaching research is that the coral is not merely a victim of its symbionts' malfunction. The host actively participates in the breakdown, because the host is trying to survive. Cells under oxidative stress do not simply endure; they respond. They turn on genes associated with detoxification, repair, and, if the damage is severe, programmed cell death. They also send signals that change the relationship with the symbiont.

A striking example comes from work by Virginia Weis's group, which used a small sea anemone model called *Aiptasia* to study cnidarian-algal symbiosis under controlled laboratory conditions. In 2006, Santiago Pérez and Weis showed that nitric oxide (NO) rises during heat stress and is implicated in the breakdown of symbiosis. Nitric oxide is a signalling molecule used throughout biology; it helps regulate blood vessels in humans, and it acts as a messenger in immune-like responses in many animals. It is also, in the wrong context, cytotoxic. In the cnidarian host, a heat-triggered surge of NO looks like an eviction notice written in chemistry. The message is not subtle: something is wrong inside, and the symbiont is now part of the problem.

There are several routes by which a coral can lose its symbionts, and it is worth keeping the language concrete. A symbiont might be digested within the host cell. It might be expelled whole, pushed out into the surrounding water. Host cells containing symbionts might detach or die and slough off. Different coral species, and even different circumstances in the same species, can favour different routes. What unifies them is not the mechanical method but the trigger: a shift from mutual benefit to net harm, mediated by oxidative stress and immune-like signalling.

A useful way to think about the trigger is as a feedback loop that flips sign. In healthy conditions, the partnership stabilises itself. Symbionts provide sugars and oxygen; the host uses those fuels to maintain the internal environment that keeps symbionts productive: controlling carbon supply, nitrogen recycling, and the conditions for calcification. Under moderate warming, metabolism speeds up. The coral's demand for energy rises, and the balance between respiration and photosynthesis can shift. If photosynthesis begins to falter while the host's energy needs increase, the partnership is strained even before acute toxicity appears.

A recent bioenergetic modelling approach by Ferdinand Pfab and A. Raine Detmer (2024) emphasises that bleaching is not simply an animal overheating like an engine without coolant. Their mechanistic picture distinguishes moderate warming, which alters nutrient exchange and energy budgets, from stronger warming, where photodamage and ROS production drive an escalating spiral. The details are complex, but the intuition is clear: the symbiont's light-harvesting apparatus becomes dangerous when it cannot process the energy it captures, and nutrient cycling can make the danger worse. Under conditions of higher dissolved inorganic nitrogen, symbionts may grow more, shifting the balance of how carbon is used and potentially increasing susceptibility to stress. Bright light, which normally powers the partnership, becomes an accelerant for oxidative chemistry.

This is part of why bleaching often appears patchy on a reef. Two colonies of the same species, a few metres apart, can bleach differently. Microhabitats matter. A slight difference in water flow can change how quickly heat dissipates and how rapidly oxygen and reactive molecules are flushed away. A colony shaded by a ledge during the brightest hours may avoid the worst photodamage even if the average temperature is the same. Corals on a reef crest experience fierce light and turbulent water; corals in a lagoon may experience calmer conditions but hotter, more stagnant water during heatwaves. The physics of the reef shapes the chemistry inside tissues.

The human face of bleaching became hard to ignore in the late twentieth century, when mass bleaching events began to be reported across oceans. One of the most widely discussed early global events occurred in 1998, coinciding with a strong El Niño. Divers and coastal communities watched familiar reefs lose their colour over weeks. In some places the corals recovered; in others, large fractions died and were overgrown by algae. For many people who lived with reefs—as fishers, tourism operators, or simply residents—bleaching stopped being an obscure scientific term and became a seasonal fear, a kind of ecological fever chart.

Scientists needed a way to translate that lived experience into something measurable and predictive. A particularly influential tool came from NOAA's Coral Reef Watch programme, which developed satellite-based heat-stress products that turn sea surface temperature into a cumulative metric called Degree Heating Weeks (DHW). The logic is deliberately practical. A single hot afternoon is not the same as weeks of elevated temperature. DHW accumulates thermal stress over roughly the previous 12 weeks, using a location-specific baseline so that "hot" means hotter than what that reef typically experiences. In the way that physicians might track a patient's fever over time rather than reading a single thermometer value, DHW tracks a heat dose.

DHW is not a chemical measurement; it does not directly count reactive oxygen species or nitric oxide. Its power is that it maps a cellular problem onto a reef-scale warning system. As a rule of thumb used in reef monitoring, bleaching risk increases once DHW reaches around 4 °C-weeks, and widespread bleaching with higher mortality risk becomes more likely around 8 °C-weeks. These numbers are not magic thresholds that apply identically everywhere, but they have become a common language between molecular physiology and management decisions. They also underscore a sobering point: bleaching is often about persistence of stress more than peak stress. The crisis builds day after day, as toxic by-products accumulate and repair systems are exhausted.

The chemical story inside a coral polyp also helps explain why small changes in global average temperature can have outsized consequences for reefs. A reef is not adapted to a fixed temperature; it is adapted to a seasonal pattern, with typical summer maxima that corals can endure because they have endured them for generations. Push the mean climate upward, and the summer peaks that used to be rare become common. Push it a bit more, and heatwaves exceed the tolerance window so frequently that recovery periods shrink. The IPCC's 2018 special report on 1.5°C made this painfully concrete: warm-water coral reefs are projected to decline very sharply between 1.5°C and 2°C of warming. The difference between those numbers can sound modest in everyday life; in coral physiology it is the difference between a stress that is sometimes survivable and one that becomes routine.

Even within a single bleaching event, the chemistry can resemble a cascade of compromises. The coral's first priority is to keep its cells alive. Oxidative stress threatens membranes; membranes are not just barriers but platforms where transporters sit, where ion gradients are maintained, where energy is handled. If membranes fail, the coral cannot regulate its internal environment. That includes its ability to manage the calcifying space where it builds skeleton. Heat stress therefore endangers not only the energy supply (because sugars from symbionts decline) but also the coral's capacity to build and maintain its limestone infrastructure. The reef's stony architecture, which took decades or centuries to grow, is maintained by a thin sheet of living tissue that is suddenly short on fuel and under chemical attack.

At this point the moral language of "divorce" becomes tempting, but it can mislead if it suggests clean separation and closure. Bleaching is messy. Some corals expel symbionts and later regain them, either from a small residual population that remained hidden in tissues or from the environment. Others recover their colour but at a cost: reduced growth, impaired reproduction, increased susceptibility to disease. Some never recover. The same act that protects a cell in the short term can weaken a colony in the long term, because the coral has traded toxicity for starvation.

A final twist is that the symbionts are not identical across all corals. Symbiodiniaceae includes multiple lineages with different tolerances to heat and light. Some symbionts are better at dissipating excess light energy as harmless heat; others maintain high productivity under a narrower range of conditions. This variation matters because the onset of toxicity depends on the symbiont's photochemistry. A coral hosting a more heat-tolerant symbiont may endure a heatwave with less ROS production, delaying or avoiding the cellular crisis that forces expulsion. But tolerance often comes with trade-offs: symbionts that are safer under heat may provide less carbon under benign conditions, leaving the coral with a smaller energy surplus in ordinary years. The partnership is tuned to a climate, and when the climate shifts, the tuning can become maladaptive.

What makes bleaching emotionally difficult to witness is that the first sign is often colour loss rather than death. A bleached coral can still extend its polyps at night, still capture plankton, still look intact from a distance. It is alive, but its internal economy has collapsed. The coral is running on emergency rations, and the reef's most reliable power source—sunlight turned into sugar—has been converted into a threat that the coral can survive only by rejecting it. A limestone city that once grew by day is suddenly paying for daylight with damage.

6.3 Evolution on a Deadline

A reef does not only die when corals die. It can die earlier, in the space between disasters, when recovery time shrinks until it becomes a mirage. The hardest part of the new reef era is not the existence of heatwaves—oceans have always had hot summers—but their repetition. Corals evolved in a world where a severe bleaching year was an exception. When exceptional becomes frequent, the biological budget stops balancing. A colony that could once afford a bad year every decade cannot afford a bad year every other year.

That rhythm matters because coral is not a fast life. A branching coral can grow quickly by coral standards and still take years to regain its former size after partial mortality. Massive boulder corals, the ones that look like old stone furniture, are slower. Meanwhile, reef fish, coastal protection, tourism, and fisheries depend on living structure, not on the memory of it. When people ask whether corals can *adapt*, they are asking whether evolution can run quickly enough to keep pace with a moving climate target.

The temptation is to imagine adaptation as a single ladder: either corals climb it or they do not. Real adaptation is more like the shifting balance of an economy with many currencies. Thermal tolerance can be purchased, but it is rarely free. It can draw down growth, reproduction, or resistance to disease. It can also depend on partnerships: a coral animal's genetics, the identity of its Symbiodiniaceae symbionts, and the wider microbial community in its tissues. Any of these can change, some more quickly than others.

One hopeful sign comes from the simple observation that not all corals bleach equally, and that this inequality has structure rather than randomness. Even during severe events, some colonies survive better than their neighbours. Over decades, reefs exposed repeatedly to heat stress can show a shift in the community toward more tolerant species, or toward more tolerant genotypes within species. Liam Lachs and Simon Donner (2023) reported an emergent increase in assemblage-level thermal tolerance on the order of a tenth of a degree Celsius per decade across several decades, consistent with "some mixture" of genetic adaptation, acclimatisation, and community turnover. A tenth of a degree is a small number in a kitchen; on a reef it is a measurable change in the temperature at which mass bleaching begins.

That phrase—"some mixture"—is not academic hedging. It is an honest description of how messy and multi-layered coral resilience can be. *Acclimatisation* means an organism changes its physiology without changing its genes: shifting the expression of heat-shock proteins, altering antioxidant capacity, adjusting membrane composition. *Adaptation*, in the strict evolutionary sense, means gene frequencies change across generations because some genotypes leave more offspring. *Community turnover* means the winners replace the losers, even if the survivors are not evolving so much as inheriting an emptier playing field. All three can raise the average tolerance of what remains. The uncomfortable catch is that a reef "getting more tolerant" can sometimes mean a reef becoming simpler: fewer species, less structural complexity, and less of the ecological richness people love.

From a conservation standpoint, the crucial question is not whether adaptation occurs in principle but whether it occurs fast enough to preserve function. Here the numbers turn harsh. The IPCC's 2018 special report on 1.5°C of warming framed coral reefs as one of the clearest examples of nonlinear risk: a projected decline of roughly 70–90% at 1.5°C, and more than 99% at 2°C. The difference between those global averages is half a degree. The difference for reefs is the gap between a world where some coral-dominated reefs still exist and a world where they become rare exceptions.

This is the deadline: not a date on a calendar, but a narrowing corridor in which evolutionary rescue might still work.

It is against that background that "assisted evolution" has become a serious phrase rather than a science-fiction one. The idea is straightforward: if natural selection is already operating, can humans help it along by identifying heat-tolerant corals and increasing their representation on reefs? The moment you try to do that, the simplicity ends. Selection is easy to invoke and difficult to steer. Heat tolerance itself is not a single trait. It can mean withstanding a short spike of extreme temperature, or enduring a long, moderately hot season without bleaching. It can mean resisting oxidative damage, or maintaining symbiont cooperation under heat-driven metabolic strain, or simply having a life history that allows rapid recovery.

Selective breeding is one of the most direct interventions, and it comes with a quiet, radical premise: corals can be treated, briefly, like crops. Researchers collect gametes during spawning events—those nights when corals release eggs and sperm in synchrony, turning the sea surface into a slick of life. Fertilisation can be done under controlled conditions. Larvae can be raised, settled onto small substrates, and grown in nurseries before outplanting.

Adriana Humanes and Liam Lachs (2024) reported that selective breeding using parents with higher heat tolerance increased tolerance in adult offspring several years later under experimental heatwave simulations. They also estimated a narrow-sense heritability for adult heat tolerance in their system around $h^2 \approx 0.2–0.3$, meaning that a meaningful fraction of variation in tolerance behaved like additive genetic variance: the kind that selection can reliably act on. This is not destiny—heritability depends on environment and population—but it is an existence proof that tolerance is not purely ephemeral physiology. There is something in the genetics worth selecting.

The same work also delivered a warning that should be printed on every optimistic press release: tolerance to short heat stress did not necessarily predict tolerance to long-duration stress. That distinction matters because oceans do not ask for permission before choosing the shape of the next heatwave. A programme that selects corals by exposing them to a brief, intense stress might produce animals that cope with a laboratory "sprint" but fail in a real-world "marathon." The genetics may be partly different; the physiology certainly is. A coral that survives a week of heat may do so by emergency triage—shutting down growth, rearranging energy use—while a coral that survives a month may need deeper stability in symbiont performance, antioxidant systems, and host control of internal chemistry.

Even if breeding works, a reef is not a greenhouse bench. Corals broadcast spawn. Their larvae drift. Genetic material flows between reefs. This raises an old worry with a dramatic name: *gene swamping*. The fear is that if you introduce heat-tolerant genotypes into a population, their advantage could be diluted by incoming larvae from less tolerant populations, preventing adaptation from taking hold. Reef-scale eco-evolutionary modelling by Yves-Marie Bozec and Arne A. S. Adam (2025), using a framework applied across thousands of

Great Barrier Reef sites, suggested that connectivity and refugia can support adaptation rather than sabotage it, and reported no evidence in their model that dispersal inevitably overwhelms selection. That does not guarantee success in the wild, but it reframes connectivity as a potential asset: movement of larvae can spread tolerant genotypes, and cooler refuges can act as reservoirs of diversity that help repopulate damaged reefs.

If breeding is one lever, symbiont choice is another. Corals do not all host the same Symbiodiniaceae lineages, and those lineages differ in their thermal sensitivity. A coral can sometimes "shuffle" its symbiont community after stress, increasing the proportion of more heat-tolerant types, though this ability varies widely across species and life stages. In theory, one could inoculate corals with more tolerant symbionts, or manipulate conditions so that tolerant symbionts become dominant.

Here again, the trade-offs sharpen. Symbionts that are safer under heat are often less generous under mild conditions. They may translocate less carbon to the host or support slower growth. The coral's limestone-building economy, so dependent on a surplus of photosynthate, can become tighter. That matters because growth is not vanity on a reef; it is the capacity to compete for space, to heal after storms, and to reproduce. A coral that survives heat but grows slowly may still lose in the long run, crowded out by algae or outcompeted by faster-growing neighbours when conditions improve.

A simple way to visualise the problem is to plot "thermal tolerance" against "growth rate" and look for the free lunch. In many systems, the free lunch is scarce.

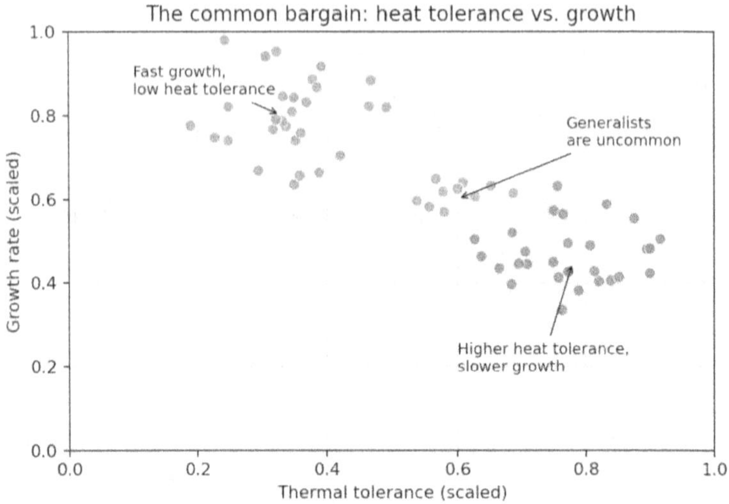

This scatter plot is only a cartoon, but it captures a recurring reality in biology: traits compete for limited resources. A symbiont that diverts energy into protective mechanisms may have less to spare for rapid growth or for exporting sugars. A host that invests in stress resistance may allocate less to skeleton-building or reproduction. The reef is built from surplus energy; resilience often spends that surplus.

Other interventions try to buy time by changing the environment rather than the biology. Shading structures can reduce light stress during heatwaves, limiting the runaway photochemistry that turns sunlight into reactive oxygen species. Localised cooling, where feasible, can blunt extreme temperatures. Reducing nutrient pollution from land can help because excessive nitrogen can destabilise the host–symbiont exchange and increase bleaching susceptibility, and because nutrient enrichment favours seaweeds that compete with corals after disturbances. These are not glamorous solutions, but they address the fact that corals face multiple stressors at once: heat, light, acidification, pollution, disease, overfishing. Removing non-climate stress can widen the margin in which adaptation has a chance.

A concrete example of how intervention becomes a public enterprise can be found in coral nurseries, which have expanded in places like the Caribbean and Florida. Fragments of threatened corals are grown on underwater "trees" or frames, tended by divers, and outplanted to damaged reefs. The technique began as a way to restore structure and preserve genotypes; it has gradually become a platform for experimentation with resilience. Which genotypes survive heat? Which grow fastest? Which resist disease? The nursery turns a reef into a managed population, with all the ethical and practical questions that implies. When a diver clips a branch of coral and ties it onto a frame, they are not only gardening; they are curating evolution, intentionally or not.

The hardest question is scale. A nursery can outplant thousands of corals. A reef contains millions. Assisted evolution may succeed locally and still be overwhelmed globally if warming continues unabated and marine heatwaves become the norm rather than the exception. Bozec and Adam's modelling work (2025) suggested that recovery becomes much more plausible if warming remains below about 2°C, in part because adaptation has more time to operate and because refugia are more likely to persist. That is a scientific way of saying something that coastal communities often express more plainly: without curbing the heat, restoration becomes an endless triage.

Still, evolution is not a romantic force that only operates over aeons. It is an accounting system that runs whenever there is heritable variation and differential survival. Corals have variation. Symbionts have variation. Reefs, as connected networks, can move variation around. The challenge is to keep enough living reef in the water long enough for that accounting system to produce something other than loss.

On a calm day, the limestone of a reef can seem permanent, like the cliff face of an island. In truth it is maintained by a thin layer of living tissue whose stability depends on a chemical partnership. When the partnership fails, the reef begins to crumble from the inside out: first the colour, then the growth, then the architecture. The most realistic hope is not to engineer a coral that never bleaches, but to keep the conditions and the diversity that allow some corals to survive, reproduce, and keep building stone in a sea that no longer keeps its promises.

Chapter 7

Six-Legged Farmers and Bodyguards

Long before humans gathered their first harvest, insects had already mastered agriculture and professional security services through sophisticated chemical engineering. By examining the antibiotic interactions of fungal gardens and the nutrient economics of ant-plants, this chapter reveals how molecular signaling transforms potential competitors into obligate allies.

7.1 The Antibiotic Gardens

A leafcutter ant trail can look, at first glance, like a moving seam in the forest floor. Each worker carries a green pennant above her head, a crescent cut from some unlucky leaf. It is tempting to imagine a simple transaction: ants harvest leaves, ants eat leaves. That is not what is happening. The leaves are cargo, not cuisine. They are a substrate, hauled underground to feed a crop.

The farmers are ants in the tribe Attini, and the best-known of them are the leafcutters, *Atta* and *Acromyrmex*. Their "fields" are not sunlit rows but dark chambers, warm and humid, packed with a pale, sponge-like fungus. The job of the workers is to keep that fungus alive, productive, and uncontaminated. The colony's wealth is not measured in the size of its leaf piles, but in the health of its fungal gardens.

Even that statement hides the real complexity. A garden is not simply an ant plus a fungus. It is a crowded, negotiated community, where cooperation has to be defended against invaders, freeloaders, and cheaters. The most striking thing about leafcutter agriculture is not that it exists, but that it survives: a monoculture, grown for years in a microbial world that specializes in invading monocultures.

The first hint that this farming system had enemies came from the gardens themselves. Healthy cultivar fungus has a distinctive look: a tufted, white-gray mat with swollen tips (gongylidia) that function as nutrient-rich packages. Ants harvest these tips like a living pantry. But sometimes, in an otherwise thriving colony, patches of the garden darken and collapse into a powdery mess. The culprit is often a parasitic mold called *Escovopsis*, a specialized pathogen that targets the ants' crop. It is not a generalist mildew that happens to wander in; it is a professional, shaped by evolution to exploit this particular agricultural niche.

Here, a basic principle of symbiosis becomes vivid. A mutualism (a partnership where both sides benefit) is not a peaceful treaty signed once and forgotten. It is an ongoing job of maintenance. The crop must be fed; the crop must be protected. And protection is expensive. In human agriculture, the cost is obvious: fences, pesticides, quarantine, and the labor of scouting. In leafcutter agriculture, it is chemistry and hygiene, deployed with an intensity that makes "ant farming" feel less like pastoral imagery and more like biosecurity.

A modern historical turning point came in the late 1990s, when Cameron Currie and colleagues reported something that field observers had long suspected but not pinned down: leafcutter ants were not merely weeding and grooming their gardens; they were carrying living antibiotic factories on their own bodies. The story begins with a simple puzzle. If *Escovopsis* is such a specialized, persistent threat, why are leafcutter gardens not constantly overrun? Ant behavior explains some of the resistance. Workers groom each other and the fungus, removing spores before they germinate. They physically excise infected sections of garden, carrying the diseased material away like hazardous waste. They adjust how they handle substrate and where they deposit it. This is all real and important. But behavior alone struggles to explain the fine-grained selectivity of control: how to suppress the parasite without poisoning the crop.

The answer involves a four-way interaction: ants, their cultivated fungus, the parasite, and bacteria that live with the ants and help police the garden. Many of these bacteria are actinobacteria, a broad group famous on a human scale because so many of our antibiotics were first discovered in actinobacterial lineages. In the ant system, the bacteria often live in specialized patches on the ants' cuticle (their outer body surface), sometimes in crypts or grooves that seem shaped to house them. The ants, in effect, wear a microbial garden on their skin.

The chemistry that follows is not a single magic bullet. It is a shifting arsenal. Different colonies can carry different bacterial strains; different strains can make different antifungal compounds; the mix of chemicals can change across geography and time. Some of the molecules characterized from ant-associated bacteria are structurally unusual, and their gene clusters can move between bacteria, allowing defensive traits to spread as if the microbiome were swapping recipes in response to a new disease pressure. This dynamism matters because *Escovopsis* is not a static enemy either. A parasite facing the same chemical weapon year after year will evolve resistance. Leafcutter agriculture survives, in part, because the defensive chemistry is not frozen in place.

The relationship is often described as "quadripartite," a term that simply means "four-part." It is worth naming the flows explicitly, because cooperation is easier to understand when it is drawn as traffic: nutrients move one way, defensive chemicals another. The ants bring leaf fragments to feed the fungus. The fungus converts plant material into edible packages that feed the ants and their larvae. The parasite attacks the fungus, exploiting the dense, predictable resource. The bacteria produce compounds that inhibit the parasite, and the ants cultivate, transport, and regulate those bacteria.

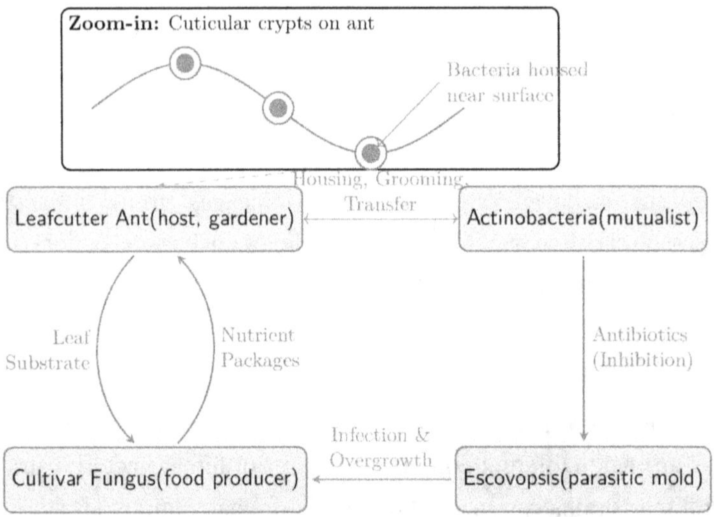

What makes this system compelling is not only that bacteria produce antifungals. Many organisms recruit antimicrobial partners. The deeper point is governance: how the ants keep the alliance working. A bacterial symbiont that produces a useful antibiotic is valuable. A bacterial symbiont that competes with other symbionts, hoards resources, or harms the cultivar is a liability. Symbiosis is not automatically stable just because it is mutualistic in principle. The partners must be chosen, maintained, and, when necessary, disciplined.

Leafcutter ants appear to manage that governance with a mixture of physical control and chemical environment. Workers constantly groom themselves and each other, distributing bacterial cells and, plausibly, the compounds they produce. Colonies can show fidelity to particular bacterial strains, and experiments have indicated that ants can discriminate between their own bacterial associates and foreign ones, even when those bacteria are close relatives. That discrimination matters in the same way that seed purity matters to a farmer: it prevents the garden from being seeded with competitors whose interests do not align with the farm.

There is also an important debate behind the scenes, and it is a healthy one, because it warns against a romantic picture of perfect coevolution. For a long time, it was tempting to imagine that ants and their bacteria had evolved together in tight lockstep over millions of years, like two species dancing in perfect synchrony. Phylogenetic reassessments have complicated that story. Evidence suggests that at least some ant lineages may acquire defensive bacteria from the environment more often than once assumed, recruiting from soil or plant material and then keeping what works. That is still coevolution in a broad sense (the ants have evolved the behaviors and body structures that make recruitment possible), but it is not the same as a single ancient bacterial lineage being faithfully inherited forever. It is closer to a managed workforce: familiar specialists may be retained, but hiring can happen when conditions change.

The parasite *Escovopsis* helps explain why such flexibility might be advantageous. Parasites specialize. When a host system is predictable, a parasite can evolve to exploit it with exquisite precision. Leafcutter gardens are predictable in several ways: the cultivar is genetically constrained relative to wild fungal communities; the nest environment is warm and humid; and the ants concentrate plant substrate in a stable location. These conditions are wonderful for fungal farming and equally wonderful for fungal pathogens. A specialized parasite can evolve to sense the cultivar's chemistry, tolerate the nest's microclimate, and spread efficiently through the dense fungal matrix.

Chemical defense, then, is not a luxury. It is a requirement for any long-lived agricultural colony. But chemical defense also carries a familiar risk: collateral damage. An antifungal compound that kills molds indiscriminately could harm the cultivar. The elegance of bacterial-produced antibiotics is that they can be deployed locally, in microdoses, and in mixtures that change the rules of the evolutionary game. In some cases, symbionts produce compounds that inhibit *Escovopsis* more strongly than they inhibit the cultivar, an important hint that the system has been tuned by selection. Sometimes the "weapon" is not a complex antibiotic at all but a change in the garden's basic chemistry, such as shifts in pH caused by microbial metabolism. Simple chemistry can be powerful if it is targeted. A pathogen adapted to one pH range may falter when the environment is pushed outside its comfort zone, while the cultivar and the ants remain functional.

The ants contribute their own chemistry as well. Many ants possess metapleural glands, paired structures that secrete antimicrobial compounds. In leafcutters these secretions can be particularly important, and experiments have shown that ants can modulate how much they deploy depending on pathogen threat. That plasticity matters because it turns defense into an adjustable expense rather than a fixed tax. If the garden is clean, constant heavy secretion would waste energy and may even disrupt beneficial microbes. If the garden is under attack, stinginess would be fatal. The ability to spend more when danger rises is, in economic terms, an investment strategy rather than a standing cost.

All of this reframes what "antibiotics" mean in nature. In human medicine, antibiotics are often imagined as a single pill with a single target, a clean intervention. In the ant colony, antimicrobial control is distributed across bodies, behaviors, and microbes. It is less like taking a drug and more like running a hospital: cleaning protocols, quarantine, specialized staff, and a pharmacy that can change its stock as pathogens evolve. The hospital analogy is not merely rhetorical. Leafcutter colonies face the same core problem that hospitals face: when you concentrate hosts (or crops) in one place, you create an environment where disease can spread quickly. The solution, in both cases, is layered defense.

A concrete example makes the point. Consider a worker returning from foraging. She carries leaf fragments that are not sterile. They come with environmental fungi and bacteria attached. Once underground, other workers take over. They chew the leaf into a pulp, shaping it into the garden matrix. During this processing, they do not simply mash leaves and walk away. They groom, lick, and manipulate the substrate, inoculating it with the cultivar and, potentially, with defensive microbes and secretions. If alien spores are present, grooming may remove them before they ever gain a foothold. If *Escovopsis* spores slip through, the ants can detect infection early, excise the infected patch, and intensify antimicrobial deployment in that zone. The important detail is locality: defense is concentrated where it is needed. That reduces harm to the cultivar and reduces selection for blanket resistance.

This locality also illuminates a subtle point about cooperation. Cooperation does not float free; it is built out of physical arrangements that make good behavior easy and bad behavior costly. The cuticular crypts that house bacteria are not just convenient storage; they are a way of controlling where bacteria live and how they are shared. A crypt is a gated neighborhood. It limits dispersal, concentrates symbionts, and may help the host regulate which strains persist. The garden chambers are similarly structured: compartmentalization can slow the spread of disease and allow targeted sanitation. Ant architecture, ant anatomy, and microbial chemistry converge on the same design principle: make the cooperative pathway robust to invasion.

There is, inevitably, a human resonance here. Leafcutter symbiosis has become a reference point in thinking about antibiotic discovery and resistance. Actinobacteria associated with ants have yielded compounds that look, to chemists, like promising leads, and their biosynthetic gene clusters reveal how nature tinkers with molecular scaffolds. Yet the more important lesson may be strategic rather than pharmacological. The ants do not rely on a single wonder drug. They rely on diversity, locality, and governance. Those are exactly the features that slow down resistance: variable cocktails instead of one compound, micro-targeted use instead of blanket exposure, and behavioral control that reduces the pathogen's opportunities.

The colony's success, then, is not a triumph of any one partner. It is the stability of a managed network. Leaves become fungus. Fungus becomes ant bodies. Parasites threaten to reroute that flow into their own reproduction. Bacteria, recruited and maintained by ants, help push the parasite back. The result is a garden that is not only an agricultural enterprise but also a living chemical negotiation, played out at the scale of spores and glands, where the price of cooperation is constant vigilance and the reward is an underground civilization fed by a domesticated microbe.

7.2 The Cost of a Standing Army

A plant cannot run. Faced with an approaching herbivore, it has only two broad options: make itself hard to eat, or recruit something that can move. The first option is familiar in the plant world. Leaves can be toughened with lignin, studded with silica, laced with bitter alkaloids, or made to ooze sticky latex. All of that is chemistry turned inward: a plant arming its own tissues.

The second option looks, at first, like outsourcing. Instead of investing entirely in thorns and toxins, a plant can pay for a security force. It offers food or housing, and in return it receives protection from animals that do have legs, jaws, and the ability to patrol. Ants are the most famous of these bodyguards because they are abundant, aggressive, and exquisitely responsive to sugar. An ant colony is, in economic terms, a standing army: always present, always hungry, and capable of mobilizing quickly when an alarm is sounded.

Standing armies are expensive. That is true for human states, and it is true for plants. If a plant pays ants with nectar and food bodies, it is diverting carbon, nitrogen, and water away from growth, roots, and seeds. If it builds housing for ants, it commits tissue that could have become leaves or flowers. Mutualism is often introduced as a win-win, but plants live under a hard budget. They cannot spend the same molecule twice.

The classic natural history case is the ant-acacia relationship described from Central America, especially involving *Vachellia* (formerly placed in *Acacia*) and ants of the genus *Pseudomyrmex*. Some acacias do not merely tolerate ants; they actively make a home for them. Their swollen thorns become hollow domes, ready-made barracks. Their leaf tips grow detachable protein- and lipid-rich packets called Beltian bodies. Along the stems they secrete sugary droplets from extrafloral nectaries, "extra-floral" meaning that the nectar is not part of pollination. It is not a bribe to bees. It is pay for soldiers.

Daniel Janzen's field experiments in the late 1960s made this system famous by showing what happens when the contract is broken. Remove the ants from a defended acacia, and the tree becomes a small, green buffet. Herbivorous insects strip leaves, browsing mammals take shoots, and even competing plants gain a foothold because the ants are no longer present to prune encroaching vegetation. The details vary by site and species, but the pattern is harshly consistent: in the right ecological setting, ant-free acacias are injured more, grow less, and survive poorly. The "standing army" is not a decorative accessory; it can be the difference between a living crown and a skeletonized shrub.

The puzzle, though, is not why plants would want protection. The puzzle is why more plants do not buy it all the time. If ants are such effective security, why is the landscape not filled with ant-housed, nectar-secreting fortresses?

The answer is cost, and cost is not an abstract moral concept. It is metabolism: the bookkeeping of molecules and energy. A plant makes sugars by photosynthesis, and those sugars are the universal currency of plant life. Sugars build cellulose walls, feed respiration, and power the assembly of every other compound. When a plant secretes sugar as nectar, it is exporting currency outside its own body. That can be worthwhile if the sugar purchases protection that prevents far larger losses to herbivores. But the bargain depends on context. If herbivores are rare, or if the ants are unreliable, the same sugar becomes a wasteful leak.

The trade-off is clearest when the rewards are not just sugar water, but expensive packages of tissue. Extrafloral nectar is already a cost: it is carbon and water delivered to the outside world. Food bodies add nitrogen and lipids, materials that can be even more limiting than carbon, especially in poor soils. Domatia, the housing structures, commit plant architecture to a particular alliance. If a plant invests heavily in these rewards and the ants fail to defend, the plant has paid for an army that never shows up.

This is where the chemistry becomes interesting, because plants do not simply open a tap and let sugar flow. They regulate, signal, and sometimes constrain access in ways that resemble contract design. The phrase "chemical contract" is not poetic fluff; it captures a real mechanism. A contract needs terms (what is exchanged), monitoring (is the other party performing), and enforcement (what happens when they do not). Plants cannot take ants to court. They must build enforcement into their biochemistry.

Start with the simplest term: nectar itself. Extrafloral nectar is usually rich in sugars such as sucrose, glucose, and fructose, but it is not chemically uniform across species, or even across times of day. The composition can shape which ants are attracted and how strongly they recruit nestmates. A small change in sugar ratio can change viscosity, evaporation rate, and palatability. Plants can also include amino acids and other minor compounds that influence ant feeding and behavior. Even without invoking any mysterious mind-control, nectar is an infochemical: a chemical signal that carries information. A well-fed ant tends to return to the source and lay pheromone trails, calling in reinforcements. The plant's secretions thus translate directly into the number of patrollers on its stems.

Monitoring is possible because ants are not subtle. When a leaf is damaged by a chewing insect, the plant's cells rupture and release a cascade of internal signals. One important class involves jasmonates, hormones central to plant defense against herbivory. These signals can induce the production of defensive compounds inside leaves, but they can also increase extrafloral nectar secretion in many species. In effect, the plant can pay more when threat rises. That is the metabolic logic of triage: spend heavily when the danger is immediate, spend less when the neighborhood is quiet.

The metaphor of an army becomes more precise here. Armies cost money even in peacetime. A permanent garrison must be housed and fed. Plants that maintain resident ant colonies in domatia pay something like a peacetime budget every day, because the ants are always present and always need food. Plants that rely on visiting ants can pay on demand, secreting nectar more strongly after damage. Both strategies have costs and benefits. A resident colony responds quickly and reliably, but it requires consistent investment and locks the plant into a partnership with a particular ant lineage. A visiting system is flexible and may be cheaper, but it risks delays and weaker defense.

Enforcement, the hardest part, is where plants become surprisingly sophisticated. Many ant-associated plants face a cheating problem. Some ants may take nectar without defending, or they may defend weakly. Other insects may steal nectar entirely, collecting rewards with no protective service. A plant can reduce theft by placing nectaries in locations that are easy for ants to patrol but harder for flying insects to exploit, or by secreting nectar in small, frequent doses rather than a large, easily stolen pool. It can also incorporate compounds that deter non-ant visitors.

In the best-studied acacias, Beltian bodies illustrate an even more pointed form of biochemical enforcement. These structures are not just nutritious; they can be chemically protected in ways that make them especially suitable for mutualist ants. Work summarized in the ant-plant literature has described protease inhibitors in some acacia food bodies. Protease inhibitors are molecules that block proteases, the enzymes that animals use to digest proteins. If a would-be thief cannot digest the protein content, the reward is less rewarding. Mutualist ants, meanwhile, may possess proteases or feeding habits that circumvent the inhibition. The result resembles a lock-and-key system: the plant's food is edible in full only to the intended partner. Even when the details vary across species, the general principle is powerful: the plant can make rewards "exclusive" not by policing behavior with a brain, but by packaging nutrients in a chemical form that favors the right consumer.

Janzen's experiments are often remembered for their drama—remove ants, watch damage explode—but they also highlight a subtler truth about cost. The acacia is not paying ants out of generosity; it is paying to avoid a larger loss. That larger loss is not measured only in missing leaf area. Leaves are factories. Each leaf is future sugar. When herbivores remove leaves, they remove photosynthetic capacity, and the plant's budget shrinks. If ants prevent herbivory, they protect the plant's income stream. Under intense herbivore pressure, paying sugar for protection can be a net profit.

Under low herbivore pressure, the arithmetic flips. If herbivores are rare, a plant might do better investing in growth and reproduction rather than feeding guards. This context dependence is one reason ant-plant mutualisms are not universal and not constant. Another reason is that ant protection is not always purely beneficial. Ants can interfere with pollinators, chase away beneficial insects, or encourage sap-feeding insects like aphids and scale insects, which produce honeydew that ants harvest. When ants farm honeydew-producing insects on a plant, the plant may experience increased pathogen transmission or resource drain. The same aggressive behavior that deters leaf-chewers can also disrupt the plant's reproductive service providers.

The trade-off can be seen at a broader scale in comparative studies. Across many ant-associated plants, there is evidence for a negative relationship between investing in direct defenses (like toxins and tough leaves) and investing in ant-mediated defenses, especially when the plant's ant rewards are costly (food bodies and domatia rather than nectar alone). This makes intuitive metabolic sense. Defensive chemistry inside a leaf costs resources, but so does exporting nectar and building housing. A plant that spends heavily on one strategy often has less to spend on the other. The interesting part is that this is not simply a choice between "be toxic" and "hire ants." Many plants mix strategies, adjusting them as conditions change, and the optimal mix can depend on soil nutrients, rainfall, the local herbivore community, and which ant species are available.

To picture the economics, imagine a young acacia seedling in a bright, disturbed patch of ground. Growth is rapid, but so is risk: tender leaves attract hungry insects. If the seedling can recruit an ant colony early, the ants can clear away herbivores and even clip back encroaching vines or neighboring seedlings, reducing competition for light. The plant's investment in nectar and food bodies is repaid by increased survival and faster growth. Now shift the scene to a shaded understory with fewer herbivores and slower growth. In that setting, constant nectar secretion might not bring enough additional benefit to offset the carbon and water costs. The same "army" that saved the seedling in open sun could become a needless expense in the shade.

It is tempting to treat these partnerships as harmonious, but the more honest view is that they are negotiated under uncertainty. Ant colonies have their own interests. They may abandon a plant if rewards decline, or if a better resource appears nearby. They may defend more aggressively when brood are present or when the colony is expanding. The plant, in turn, may adjust its rewards in response to damage, season, or internal reserves. This is not sentiment; it is feedback. Nectar secretion can rise after herbivore attack and fall when the threat subsides. Food body production can vary with leaf flush. Even domatia can be produced more or less depending on growth stage. The partnership persists because, most of the time, the feedback keeps both sides within a range where cooperation pays.

A useful way to connect this back to chemical cooperation is to notice what the plant is actually purchasing. It is not purchasing "antness." It is purchasing behavior: patrolling, biting, stinging, pruning, recruiting nestmates, and responding to alarm cues. Behavior is an output of physiology, and physiology responds to chemicals. Sugar intake fuels activity. Certain nectar constituents can affect feeding and recruitment. Plant damage signals can alter nectar flow and perhaps the cues that ants use to search. The plant's budget is paid in molecules, and the defense is delivered in actions.

There is also a moral hazard, familiar from human economics. If a plant relies heavily on ant defenders, it might reduce its own direct defenses, making itself more vulnerable if ants disappear. That can create a trap: a plant specialized for ant protection may do poorly without ants, even if ant presence is not guaranteed. The ant-acacia system shows how extreme that specialization can become. A well-defended acacia may have relatively weak physical defenses compared to a thorny plant that relies on itself, because the ants are expected to do the fighting. Remove the ants, and the plant's vulnerability is exposed.

One can see why Janzen's work struck such a chord. It was not merely a charming story about ants living in thorns. It was a demonstration that a partnership can be so integrated that removing one partner changes the other's fate dramatically. That is a hallmark of deep mutualism: not a casual exchange, but a reorganization of life history around a trade.

The most revealing detail, perhaps, is that the acacia is not feeding ants with "spare sugar." There is no such thing as spare sugar in a plant struggling for light, water, and nutrients. Every droplet of nectar represents an opportunity cost: a fraction of growth not taken, a fraction of seed not produced, a fraction of root not extended. The standing army is paid from the same account that funds reproduction.

And yet the payment continues, because herbivores are relentless, and because deterrence works best when it is credible. A plant that only occasionally hires guards may be tested constantly by insects that can sense vulnerability. A plant that reliably supports an aggressive ant colony builds a reputation in the local ecological neighborhood. Herbivores learn, in their own way, to avoid the leaves that bite back. The plant's chemistry purchases not only immediate defense but a sustained threat, maintained on sugar, written into the daily traffic of ants along its stems.

7.3 Molecular Seduction and Fidelity

A flower is a chemical broadcast station. Long before a bee arrives, molecules are already leaving the petals and drifting into the air, forming an invisible plume that widens, thins, and bends with every eddy of wind. To a human walker, that plume may register as "sweet" or "spicy" or not at all. To an insect, it can be as legible as a road sign: not only *food is here*, but *which food*, *how much*, and sometimes *who else is visiting*.

That last point matters because pollination is a peculiar kind of cooperation. Ant bodyguards and bacterial antibiotics can be paid for and, to some degree, managed. Pollinators are different. They come and go. They have their own schedules and preferences. A plant's problem is not simply to attract an insect, but to attract the *right* insect, often with exquisite specificity, and to do so repeatedly across days and seasons. A flower is a temporary organ built for a narrow window of opportunity. If pollen is not moved to another plant of the same species, the entire investment is wasted.

The chemistry that mediates this exchange is dominated by volatile organic compounds, or VOCs. "Volatile" means they evaporate easily and travel through air; "organic" here means carbon-based. A single flower can release dozens or hundreds of VOCs: terpenes (built from isoprene units, often citrusy or resinous), benzenoids (aromatic ring compounds, often sweet or spicy), and nitrogen- or sulfur-containing compounds that can be pungent, even shocking. The bouquet is rarely accidental. It is shaped by enzymes whose job is to assemble and modify small molecules into signals that insects can detect at astonishingly low concentrations.

The insect, for its part, is not passively "attracted" the way iron filings are attracted to a magnet. It is reading chemistry with an organ we can barely imagine: antennae packed with sensory hairs, each hair riddled with pores, each pore admitting molecules that bind to odorant receptors. Binding is the crucial step. A receptor is a protein shaped to be more compatible with some molecules than others, like a lock that accepts certain keys. When a VOC docks, it triggers electrical activity that the insect brain interprets as a smell. The insect's behavior then follows a learned or innate script: fly upwind, land, probe, feed, search, leave. A plant that can reliably press the right buttons in that sensory system can steer insect movement as effectively as if it had tiny reins.

The seduction, though, is not necessarily honest. Many flowers pay pollinators with nectar, a sugary solution that fuels flight. Some provide pollen, rich in protein and lipids. In those cases the scent is, broadly, advertising: a signal correlated with a reward. But other plants cheat. They emit scents that mimic food, mates, or safe places to lay eggs, and the pollinator receives little or nothing. The plant still gets pollen moved. The insect gets tricked. This is not an exception at the fringes of botany; it is a recurring evolutionary strategy.

A concrete example is the orchid genus *Ophrys*, often called the bee orchids. Some *Ophrys* species attract male bees or wasps by emitting VOC blends that mimic the sex pheromones of females. The male, detecting what his brain interprets as a receptive mate, attempts to copulate with the flower. In doing so, he picks up pollen packets (pollinia) and transfers them to the next deceptive bloom. The "reward" is not nectar but a burst of misplaced effort. The orchid's success depends on chemical mimicry so precise that males are willing to be fooled repeatedly, at least until learning or exhaustion intervenes.

The other pole of the spectrum is honest mutualism, where chemistry is used not to deceive but to build fidelity. Fidelity means repeat business: the same pollinator returning to the same plant species, ignoring competitors even when other flowers are nearby. Fidelity is valuable because pollen is not a generic substance. Pollen delivered to the wrong species is wasted, and in some cases it can even clog stigmas or interfere with fertilization. A plant benefits when its chemical signals reliably recruit the insects most likely to deliver pollen to conspecifics.

Peter Raven, writing in the 1960s and beyond, helped popularize a broad coevolutionary view of plant—insect interactions: plants evolve chemical and structural traits, insects evolve counter-traits, and diversification can accelerate as lineages specialize. Pollination is one of the cleanest arenas for this, because it often involves tight pairing between floral traits and insect sensory systems. A mutation that tweaks scent composition may shift which pollinator species visits most. If that shift reduces gene flow with the ancestral pollinator set, it can contribute to reproductive isolation, the early scaffolding of new species. The remarkable thing is that this can happen without any dramatic physical barrier. A slight change in the "smell key" and the "receptor lock" can rewire a plant's mating network.

What does it mean, mechanistically, for a plant to tweak a scent? In biochemical terms it can be as small as a single enzyme variant. Many floral VOCs are synthesized through pathways that branch like city streets. Change the activity of an enzyme at one junction, and the traffic of metabolites is diverted. More of compound A, less of compound B. Add a methyl group here, oxidize there. These are tiny edits at the molecular scale, but they can be enormous edits at the ecological scale if the pollinator's receptors are tuned sharply.

Pollinators are not blank slates, either. Their receptors evolve under selection to find food efficiently and mates reliably. If a pollinator lineage begins to rely heavily on a particular plant's nectar, individuals whose receptors better detect that plant's bouquet gain an advantage. They arrive sooner, feed more, raise more offspring. Over generations, receptor variants that improve detection become more common. The plant's scent and the insect's sensory system can thus ratchet together: a feedback loop where each side's evolution shapes the other's environment.

This is the alluring, almost romantic version of coevolution: flowers and insects locked in a tight embrace. Real systems are messier. Plants often have multiple pollinators; insects often visit multiple plants. Weather and season shift the terms of trade. Yet even in messy systems, chemistry allows plants to bias probabilities. A flower cannot command a bee, but it can stack the deck by producing a bouquet that certain bees are primed to notice, by timing emission to match the pollinator's daily activity, and by linking scent to reward so that learning reinforces loyalty.

Learning is crucial, and it is often underappreciated when pollination is described as pure instinct. Many bees learn floral odors quickly and can associate a particular scent profile with a profitable nectar source. That learning can make honest signals more powerful: a plant that reliably rewards a bee trains the bee's nervous system to prioritize its odor. Fidelity emerges not from a binding contract but from a feedback between plant chemistry and insect memory.

Deception has its own constraints. A plant that cheats too aggressively risks being abandoned as pollinators learn. Deceptive orchids often avoid this by being rare relative to rewarding flowers, or by flowering in patterns that reduce repeated encounters by the same individual. Some exploit sensory biases that are hard to unlearn, such as sexual cues. Even then, deception is a high-wire act, and it helps explain why many plants settle into a middle ground: offering a modest reward while also manipulating behavior through scent architecture.

Behavioral manipulation can be subtle. Consider how a scent plume works in moving air. Insects often navigate upwind by "casting," zigzagging through odor filaments to stay within the plume. A plant that emits stronger scent can be detected from farther away, but that emission is not free. It costs carbon and energy, and it can attract unwanted visitors: nectar thieves, florivores that eat petals, or predators that ambush pollinators. The plant must balance reach against risk, a chemical version of turning up the volume in a dangerous neighborhood.

The idea of risk connects pollination back to protection, and not only in metaphor. Ants, which can be excellent defenders of leaves, are often poor companions for flowers. They can chase away or attack pollinators, and their presence can contaminate floral surfaces. Some plants therefore face a conflict between recruiting ants for defense and recruiting insects for pollination. One way plants manage this is spatial separation: extrafloral nectar on stems and leaves, floral nectar deep within flowers. Another is chemical separation. Floral scents and nectar chemistry can be tuned for flying pollinators, while extrafloral nectar can be tuned for ants, sometimes with compounds that suit one group and deter the other. The plant's surface becomes partitioned into chemical zones, each with different intended customers.

That partitioning is, again, contract design. A plant cannot prevent every opportunist from taking a reward, but it can make some rewards more accessible to the partner that provides the desired service.

Extrafloral nectar can be secreted in small pulses that encourage ant patrol without creating a large pool for other insects. Food bodies can be packaged with biochemical "locks" that mutualist ants can open more efficiently than exploiters. In pollination, the analogous locks are often sensory rather than digestive: a scent blend that a particular pollinator recognizes as *its* flower, even when other blooms are nearby.

There is a striking parallel here with microbial fidelity in ant gardens. In the leafcutter system, ants can discriminate between their resident bacterial symbionts and foreign strains, a form of partner choice that keeps the defensive microbiome coherent. In flowers, the plant is attempting something similar but with different tools. It cannot physically exclude "foreign" insects in the same way, but it can bias the visitor community by making its signals more legible to some species than others, and by rewarding the visitors whose bodies are most effective pollen couriers.

The most dramatic cases of specificity often involve specialized pollinators with matching morphology: long-tongued moths and long-spurred flowers, for instance. But chemistry can be equally specific even when the shapes look ordinary. Two closely related plant species may have similar petals and similar nectar volumes, yet emit different ratios of the same VOCs. To human noses they might smell equally sweet. To an insect, the difference can be categorical. A tiny shift in ratio can flip a receptor's activation pattern from "ignore" to "approach," because insect olfaction is combinatorial: identity is encoded by the pattern across many receptors, not by a single molecule.

A historical anecdote helps ground this abstract talk of receptors and ratios. Long before the molecular details were known, naturalists were struck by the way certain flowers seemed "made" for certain insects. Charles Darwin's work on orchid pollination in the 1860s is a classic example. He showed that orchids often have intricate mechanisms that force pollinators into specific movements, attaching pollen precisely. Darwin's genius was to see these mechanisms as products of natural selection rather than design. What is sometimes missed is that orchids do not rely only on mechanical fit. They also rely on chemical persuasion. Darwin did not have gas chromatography to analyze VOCs, but the logic of his observations fits seamlessly with modern chemistry: the flower must bring the right animal to the right place on the right day, and chemistry is the long-distance invitation that makes the mechanical trap relevant.

The chemical side of pollination also has very practical human echoes. Orchard productivity depends on pollinator behavior; crop breeders sometimes inadvertently alter floral scent while selecting for yield, changing pollinator visitation. Conservationists worry about pollination networks disrupted by habitat loss, pesticides, and climate-driven shifts in flowering time. In those contexts, floral VOCs are not just charming fragrances. They are part of the infrastructure of reproduction, and slight disruptions can propagate through ecosystems.

Climate and environment can change the signals themselves. VOC emission depends on temperature, humidity, and plant health. Heat can increase volatility and emission rate, potentially enlarging the plume but also exhausting reserves faster. Drought can reduce nectar production and alter the composition of both nectar and scent. A pollinator that has learned that a particular odor predicts a good reward may find, in a drought year, that the promise is broken. Fidelity can unravel when the chemical link between signal and reward weakens.

Even in stable conditions, fidelity is never perfect. It is a probability distribution. The plant's goal is to tilt that distribution toward effective pollinators, just as an ant-associated plant tilts its reward distribution toward ants that actually defend. When the tilt is strong, specialization emerges. When it is weak, generalism persists. Both can work, depending on ecological context.

The most revealing insight, perhaps, is that plants do not merely *attract* insects; they shape insect decision-making in a way that makes cooperation plausible without trust. In the language of economics, the plant reduces search costs for the pollinator by making itself easy to find. In the language of neurobiology, it triggers sensory circuits that bias approach and feeding. In the language of chemistry, it emits a blend that docks to receptors with enough specificity to stand out from background odors. These are different vocabularies for the same phenomenon: molecules leaving a flower and rewriting the behavior of an animal.

A flower's fragrance can feel ephemeral, even frivolous, because it disappears as soon as the breeze shifts. Yet for a plant, that fragrance is a reproductive strategy made of carbon atoms, released with purpose, calibrated by selection, and interpreted by a nervous system that evolved to treat certain molecules as instructions. The pollen does not move itself. The plant cannot chase its mate. It can only persuade the world to do the moving, and it does so one volatile molecule at a time.

Chapter 8

The Multitude Within

We often conceive of ourselves as singular individuals, yet biologically we are walking ecosystems teeming with trillions of foreign cells. This chapter peels back the skin to reveal how the chemical whispers of our gut inhabitants dictate everything from metabolic regulation to emotional states, effectively blurring the line between host and passenger.

8.1 The Chemical Holobiont

A human body looks like a single thing from the outside: one skin, one name, one set of fingerprints. Open the view a little, and the singularity softens. The surface is crowded with microbial life. The mouth has its own distinctive residents. The gut holds a vast, restless population whose collective chemistry never entirely switches off. The older picture of the individual organism as a sealed fortress begins to feel like a useful fiction. A better metaphor is a negotiated border, maintained not just by immune vigilance but by trade.

Biologists use the word *holobiont* for this kind of composite being: a host and the many microorganisms that live on and in it, taken together as an ecological unit. The term does not imply harmony at every moment, and it does not mean the microbes are merely "organs" we own. It points to a stubborn fact: much of what a body does is done in partnership, and that partnership is chemical. Seth Bordenstein and Kevin Theis, writing in 2015, argued that thinking about animals without their microbial associates leaves out fundamental principles of how organisms work. Eugene Rosenberg and colleagues had earlier pushed the stronger idea that host plus microbiome might sometimes behave as a unit in evolution. That claim has been debated, and it deserves the kind of precision that evolutionary biologists prize: selection can act at multiple levels, and not every microbial passenger is a coevolved partner. Yet the everyday physiology is less controversial. Your gut lining, your immune system, and the microbes in your intestine are locked into continuous metabolic exchange, and the "you" that results is not purely human chemistry.

The surprise is not that microbes exist in the gut; the surprise is how deeply their molecules are woven into what we call normal. The gut is an extreme environment: low oxygen, periodic floods of food, waves of acid and bile upstream, and an immune system poised to react. If the intestine were merely a tube for absorbing calories, it might tolerate some microbial freeloading. Instead it behaves like a managed habitat: it feeds microbes, corrals them, listens to their metabolites, and punishes those that get too ambitious.

The first and most underestimated piece of feeding is mucus. The colon is lined with a slippery, structured layer largely built from mucins, huge glycoproteins made by the host. It is easy to think of mucus as a simple barrier, a kind of biological plastic wrap. In reality it is also a ration. Mucins carry complex sugars and chemical decorations; for some microbes, those sugars are dinner. The host produces them in abundance, and in doing so it shapes the microbial community. When dietary fiber is scarce, many bacteria turn to mucus-derived carbohydrates; in a sense, the body becomes the pantry. That is not a comforting thought, but it is a revealing one. The host is not only defending against microbes; it is provisioning them in a controlled way, which is exactly what one would do in a long-term relationship where eviction is impossible.

In exchange, the microbes deliver small molecules that the host cannot easily make on its own. Among the most celebrated are short-chain fatty acids (SCFAs): acetate, propionate, and butyrate. These are produced when gut microbes ferment dietary fiber and other substrates that human digestive enzymes leave behind. The chemistry is simple enough to sketch on a napkin, yet the consequences are broad. SCFAs are fuels, signals, and sometimes epigenetic nudges, altering which genes get expressed by influencing histone deacetylases and receptors on host cells. Ara Koh and colleagues, in a 2016 synthesis, described SCFAs as key mediators translating diet into host physiology.

Butyrate has acquired a special status in this story because it offers an unusually direct example of metabolic interdependence. Colonocytes, the cells lining the colon, use butyrate as a major energy source. The host builds a habitat; microbes thrive in it and return a fuel that keeps the habitat healthy. This feedback loop is not sentimental. It is closer to the logic of a city that pays for its water system by taxing the factories that depend on it. If the factories disappear, the city's infrastructure falters; if the city neglects the pipes, the factories cannot function. Cooperation, in living systems, is often a matter of reciprocal dependence maintained by consequences.

The immune system is part of this economy. It is sometimes taught as an army trained to kill invaders, but in the gut it behaves more like customs and immigration. Most microbes in the lumen are not eliminated. Instead the immune system decides where they may go, how close they may approach, and which behaviors will trigger force. A key tool is immunoglobulin A (IgA), an antibody secreted into the gut that can coat bacteria and limit their ability to attach to the epithelium. Another is the production of antimicrobial peptides. Yet even these "weapons" often act as regulators rather than indiscriminate bombs. The gut's problem is not "microbes" in general; the gut's problem is microbes in the wrong place, in the wrong quantity, or expressing the wrong genes at the wrong time.

Here SCFAs and other microbial metabolites become a kind of molecular paperwork. Immune cells have receptors that detect them. Host cells respond by strengthening barrier function, tuning inflammatory thresholds, and encouraging the development of regulatory T cells that keep immune responses from becoming unnecessarily destructive. The details matter: an immune system trained in a sterile environment can be jumpy, more prone to react to harmless stimuli. An immune system trained amid microbial diversity learns discrimination. This is one reason why modern discussions of allergy and inflammatory disease so often circle back to microbial exposure and microbial metabolites. The immune system is not merely suppressing microbes; it is learning from them.

There is a historical reason this idea took time to become respectable. For much of the nineteenth and early twentieth centuries, microbes were primarily framed as enemies. That framing was not irrational. Cholera, tuberculosis, puerperal fever, and a host of other infections made the microbial world a mortal threat. Louis Pasteur's demonstrations that microorganisms drive fermentation, and Robert Koch's formulation of postulates linking specific microbes to specific diseases, transformed medicine. The germ theory was a triumph precisely because it was clear: identify the culprit, eliminate it, save the patient.

Yet even in that era, there were hints that microbes were not only villains. Ilya Metchnikoff, a Russian-born immunologist working at the Pasteur Institute, won the Nobel Prize in 1908 for work on phagocytosis, a cornerstone of immune defense. He was also fascinated by longevity and diet, and he speculated that fermented milk containing "useful" bacteria might improve health by suppressing putrefactive microbes in the gut. Some of his claims read today as overly confident, even quaint, but his instinct that microbial life could be harnessed for benefit was prescient. Metchnikoff was imagining an inner ecology before "ecology" became a common framework for the body.

The holobiont concept inherits both legacies: the fear of infection and the recognition of mutualism. It refuses to choose between them, because the gut does not choose. A microbe can be a helpful fermenter in one context and a dangerous opportunist in another. *Escherichia coli* is an instructive example. Many strains are ordinary residents of the human intestine, and some contribute to vitamin production and colonization resistance. Other strains carry virulence factors and can cause severe disease. The difference is genetic, ecological, and situational. The holobiont is not a utopia; it is an arena where cooperation and conflict are managed by chemistry and constraints.

Thinking in terms of management raises a difficult question: why does the host tolerate a dense microbial population at all? Why not keep the gut as sterile as possible? The shortest answer is that the host cannot extract all potential nutrition alone, and the cost of keeping the gut sterile would be enormous and perhaps impossible. Plant cell walls and many complex carbohydrates pass through the small intestine largely intact. Microbes can break them down, recovering energy that would otherwise be wasted, and converting it into SCFAs that the host can absorb. There is also the matter of competition. A gut packed with benign residents leaves less room for pathogens, a phenomenon known as colonization resistance. The host, in effect, hires microbial tenants to keep out worse squatters.

This "hiring" is not done by conscious choice, of course. It is done by selection at multiple scales: across evolutionary time, and within a lifetime. Within a lifetime, the host shapes the gut habitat through pH, bile acids, mucus, and immune signaling. Microbes respond by adapting, competing, and exchanging genes. The result is a dynamic community whose membership can change quickly while its overall functions may remain more stable. That distinction-membership versus function-matters when people speak casually of "the microbiome" as though it were a fixed organ. It is more like a workforce: individuals come and go, but the work continues as long as the conditions and incentives are in place.

The gut lining is where the chemical negotiations become most visible. Imagine the colon not as an empty tunnel but as a crowded shoreline. On one side is the microbial sea, dense with cells, enzymes, and metabolites. On the other side are the host tissues, richly vascularized, immunologically active, and desperate to avoid inflammation. Between them lies a thin boundary of mucus and epithelial cells, with tight junctions acting as molecular rivets. That boundary is not a wall; it is a membrane. Molecules cross it in both directions, and the nature of those molecules determines whether the relationship feels like peace or siege.

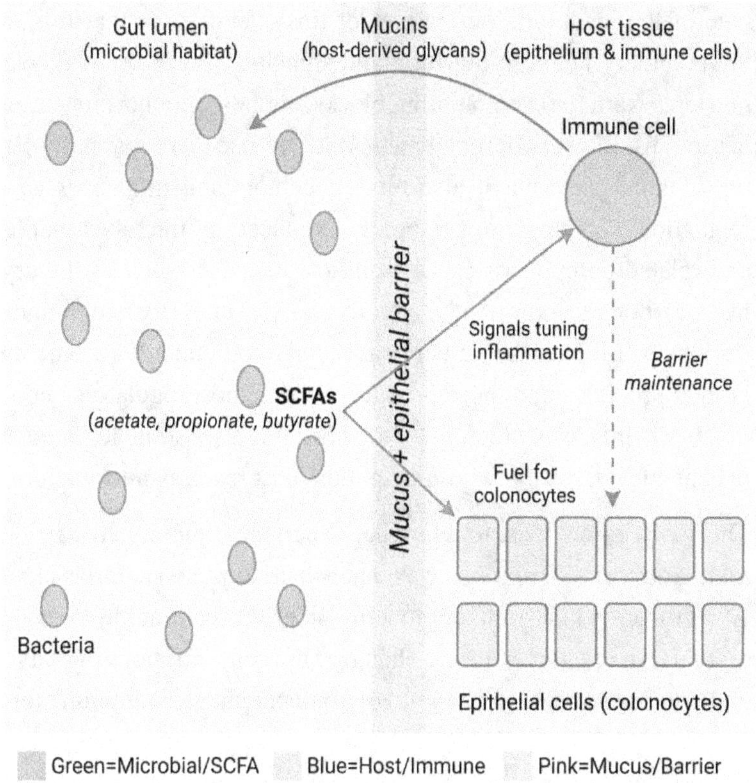

The diagram is almost too tidy. Real guts are messier: mucus comes in layers; microbial species are thousands, not a handful; and the immune system has a richer vocabulary than a single arrow. Yet the central exchange is accurate. Host-derived molecules feed microbes; microbial metabolites feed and educate the host. The term "educate" can sound mystical, but it is simply physiology repeated until it becomes baseline. A developing immune system exposed to microbial metabolites calibrates its thresholds. Epithelial cells bathing in butyrate adjust gene expression accordingly. Over time, the host begins to assume the presence of these chemicals, and what we call "health" may depend on that assumption.

One of the most striking features of this chemistry is that it blurs the boundary between nutrition and signaling. In older textbooks, nutrients were fuel and building blocks; signals were hormones and neurotransmitters. Microbial metabolites refuse that separation. Butyrate is calories, but it also binds receptors and influences gene regulation. Tryptophan metabolites produced by microbes-indoles and related compounds-can act on host receptors such as the aryl hydrocarbon receptor (AhR), affecting barrier integrity and immune responses. Molecules that begin as dietary components become microbial products, and those products become host regulators. Food, in other words, is not merely for the host; it is a substrate for an internal chemical industry, whose output the host reads as information.

This gives a new twist to a familiar experience: dietary change can make you feel different in a way that is hard to pin on calories alone. A shift from a fiber-rich diet to a low-fiber one can rapidly alter microbial fermentation patterns, changing the concentrations of SCFAs and other metabolites. The gut environment changes; immune tone changes; motility can change. It is not that microbes are puppeteers and humans are puppets. It is that the host is a metabolic landscape, and landscapes are altered by what flows into them.

Once you accept the body as a chemical marketplace, some medical puzzles look less mysterious. Consider why antibiotics, while lifesaving, can be physiologically disruptive beyond their intended targets. Killing bacteria in the gut is not like clearing weeds from a garden path; it is closer to shutting down a factory district. The supply of certain metabolites collapses, and opportunistic organisms can rush into the vacuum. *Clostridioides difficile* infection is the notorious example: after antibiotics, the normal community that suppresses it can be diminished, allowing a toxin-producing pathogen to proliferate. Here again the holobiont framing helps. The "patient" is not only a human tissue under assault; it is a whole ecosystem thrown out of balance.

The holobiont perspective also explains why simple moral language-"good bacteria, bad bacteria"-rarely survives contact with real data. A microbe that produces helpful SCFAs in one person might contribute to gas and discomfort in another, depending on diet, gut transit time, and the rest of the community. A bacterium that is benign in the gut can become dangerous if it crosses the intestinal barrier into the bloodstream. Even the host's behavior matters. Stress hormones can alter gut motility and secretion; sleep patterns affect metabolism and immune function; these changes ripple outward into microbial ecology. Cooperation in biology is often conditional.

Conditional cooperation raises the question of enforcement. How does the host prevent microbes from taking the resources and giving nothing back? There is no police force in the literal sense, but there are mechanisms that resemble sanctions. The host can change the availability of oxygen and nitrate in the gut, shifting which microbes thrive. It can secrete antimicrobial peptides selectively. It can increase mucus production, changing habitat structure. It can deploy IgA that targets particular bacterial traits. These are not conscious strategies, but they function as ecological constraints. Microbes that behave in ways that damage the host risk provoking an inflammatory response that makes the environment hostile. In extreme cases, the host dies, and the microbes lose their habitat. Many microbial lineages, over evolutionary time, have been filtered by this simple arithmetic: those that keep the host alive keep their home.

At the same time, microbes have their own leverage. Their metabolites can encourage tolerance; their presence can outcompete pathogens; their fermentation can supply energy. The host benefits from stability and will often "prefer" a community that provides it, in the loose sense that host physiology will tend to reinforce that state. In ecology, feedback loops can stabilize a system without any planner. The gut is full of such loops.

There is an emotional sting in this view of the self. We tend to want individuality to be clean: my body, my brain, my choices. The holobiont does not deny agency, but it complicates the story. The chemicals that help maintain your gut barrier and train your immune system are, in many cases, made by organisms that are not you in any genetic sense. The "within" is not a metaphor; it is a population. The blood that leaves your gut after a meal carries not only glucose and amino acids but also a shifting mixture of microbial products and host responses. You are, in part, the downstream effect of other metabolisms.

That complexity does not make the individual meaningless. It makes the individual more realistic. The unit that walks around, signs forms, falls in love, and gets sick is still one person. Yet that person is held together by negotiated chemistry: a border that is continually rebuilt, a metabolism that is partly outsourced, and an immune system trained by the molecules of longtime residents. When cooperation works well, it becomes invisible, the way breathing is invisible until it is difficult. When it falters, the body reminds you that individuality was never a matter of isolation, but of successful coexistence.

8.2 The Psychobiotic Highway

Mood feels private. Anxiety can rise in a quiet room with no obvious cause; confidence can evaporate between breakfast and lunch. Because these shifts arrive as thoughts and sensations, it is tempting to locate their origin entirely in the brain, as if the skull were a sealed command center and the rest of the body merely carried out orders. Yet the brain is bathed in blood, supervised by immune cells, and wired to the organs that keep it alive. The gut, in particular, is not a distant supplier of calories but a chemically active interface where microbial metabolism, epithelial signaling, and immune decisions are made continuously. When people speak of a "gut feeling," the phrase turns out to be less poetic than anatomical.

The gut—brain axis is the name given to this two-way traffic: signals going from brain to gut (stress changing motility, secretion, appetite), and signals going from gut to brain (metabolites, hormones, immune mediators, and neural impulses shifting arousal, pain, and sometimes behavior). The difficulty is not imagining communication; it is resisting a too-neat story in which microbes "make serotonin" and therefore directly manufacture happiness. Biology seldom grants such shortcuts. What microbes can do, with impressive chemical fluency, is alter the context in which the nervous system operates: the tone of inflammation, the permeability of the gut barrier, the set points of stress circuits, the availability of precursors for neurotransmitters, and the firing patterns of gut-connected nerves. These are not small perturbations. They are the conditions that make one mental state easier to enter than another.

A useful way to picture the axis is as three parallel lanes that often carry the same message in different dialects. One lane is neural: rapid, wired, and usually precise. Another is immune: slower, more diffuse, but powerful, because inflammatory signals reshape whole-body physiology. The third is chemical through the bloodstream: metabolites and hormones that circulate widely, with effects that depend on which receptors are present in a given tissue. The same gut event can activate all three at once. A meal changes microbial fermentation; fermentation changes metabolites; metabolites act on gut endocrine cells and immune cells; immune signals and hormones enter blood; and sensory neurons in the gut report upstream.

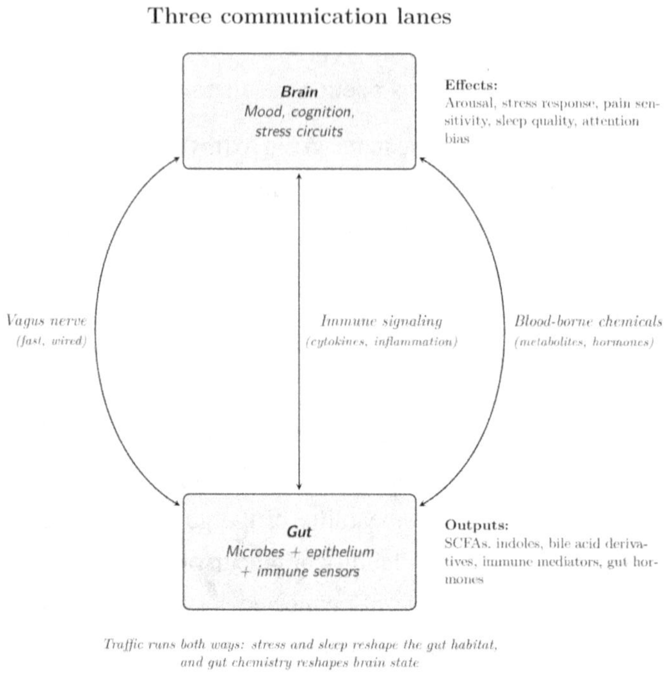

The first lane, the vagus nerve, is the celebrity route. It is the largest of the cranial nerves, and it threads through the torso with branches to the heart, lungs, and much of the digestive tract. It carries motor commands from brain to gut, but it also carries sensory information upward: stretch, tension, irritation, and chemical context. If you have ever felt nausea arrive as a whole-body verdict rather than a localized stomach complaint, you have experienced vagal signaling as a kind of embodied certainty.

What gives the vagus nerve its special place in microbiome discussions is that it offers a plausible mechanism for speed. The immune system can take hours to shift; blood-borne metabolites can circulate quickly but may be buffered or filtered; a nerve impulse is immediate. In 2011, Javier Bravo, Paul Forsythe and colleagues reported a striking result in mice: feeding a particular *Lactobacillus* strain altered stress-related behavior and changed expression of certain GABA receptors in the brain, and these effects disappeared when the vagus nerve was severed. It was not a human trial, and it was not a promise of a probiotic cure for anxiety. It was something rarer and more valuable: a demonstration that gut microbes could, under controlled conditions, influence central neurochemistry through a defined neural pathway.

Even that result invites restraint. A mouse is not a small human with whiskers; it is a different animal with a different gut anatomy, different microbiota, and a brain shaped by different evolutionary pressures. Still, the experiment shifted the conversation. Before such work, many claims about gut microbes and mood floated in the fog of correlation: people with depression have different microbiome patterns, children with autism show differences, stress reshapes microbial diversity. Correlation can be a clue, but it is not a mechanism, and it is not causation. A vagus-dependent effect pins at least one thread of causality to a physical wire.

The second lane is immune signaling, and it may be the most underestimated when people are imagining "mental" life. The brain is often described as immunologically privileged, but it is not immunologically isolated. Inflammatory cytokines in the bloodstream can alter how the brain functions; microglia, the brain's resident immune cells, respond to systemic signals; and the blood—brain barrier itself is dynamic, changing permeability under certain conditions. A chronically activated immune system can produce "sickness behavior": fatigue, social withdrawal, reduced appetite, and a kind of cognitive slowing that looks, from the inside, like low mood. Anyone who has had influenza recognizes the mental signature of inflammation: it is hard to feel optimistic when the body is broadcasting alarm signals.

Gut microbes have multiple ways to influence immune tone. They can support barrier integrity, keeping bacterial components and inflammatory triggers from leaking into the bloodstream. They can produce metabolites that dampen or shape immune responses. They can also, when the community is disrupted, contribute to conditions that favor inflammation. The gut lining is not simply a passive filter; it is an immune organ. When it is irritated or permeable, immune cells receive more stimulus and can respond with cytokines that circulate. This is not a claim that all depression is "inflammation" or that all inflammation causes depression. It is a recognition that immune signaling is a route by which gut ecology can influence the brain without any microbe needing to colonize neural tissue or "send thoughts."

The third lane, blood-borne chemical signaling, is where the popular story about serotonin usually appears. The fact at the core is correct: a large proportion of the body's serotonin is produced in the gut, largely by enterochromaffin cells, and the gut's microbial community can influence serotonin-related physiology. But the implication commonly drawn is misleading: serotonin made in the gut does not simply cross into the brain and become mood. Serotonin itself does not freely pass the blood—brain barrier, and brain serotonin is primarily synthesized within the nervous system from the amino acid tryptophan. The gut's serotonin is more directly involved in intestinal motility, secretion, and local signaling.

Yet the gut can still influence the brain's serotonergic system indirectly, in ways that are chemically plausible and experimentally supported in various models. Microbes can alter the availability of tryptophan, either by consuming it, transforming it into other compounds, or shifting host metabolism in ways that change how tryptophan is routed. Microbes can produce a family of tryptophan-derived metabolites-indoles and related molecules-that act on host receptors such as AhR, shaping barrier function and immune signaling. In

2020, researchers showed that certain microbial tryptophan metabolites could protect barrier integrity via AhR pathways in a model of colitis-associated permeability, a reminder that a metabolite can affect the brain by first affecting the gut wall and the immune system that monitors it.

The blood lane also carries short-chain fatty acids, bile acid derivatives, and other microbial products that can bind receptors on cells throughout the body. Some of these receptors are expressed on immune cells and on the vagal afferents that relay information to the brainstem. Others are present on endocrine cells that release hormones into circulation. The signal is rarely a single molecule; it is a chemical "weather pattern," a shifting distribution of metabolites and mediators that change the probability of certain physiological states.

At this point, someone usually asks a simple question: if microbes can manufacture neurotransmitters-serotonin, dopamine, GABA, acetylcholine-why not treat mental illness by changing the microbes? Part of the answer is that the words are doing too much work. A bacterium can synthesize a molecule that is chemically identical to a human neurotransmitter, but where that molecule acts depends on location, concentration, receptors, and barriers. GABA produced in the gut is not automatically GABA in the brain. Dopamine in the intestinal lumen is not dopamine bathing the prefrontal cortex. The chemistry is real; the leap from chemical identity to psychological effect is where most hype lives.

The more interesting story is less direct and, because it is less direct, often more believable. Gut microbes can modulate the enteric nervous system, the dense network of neurons embedded in the gut wall that coordinates digestion and communicates with the central nervous system. They can influence gut motility and sensitivity, which feeds back into brain circuits processing discomfort, vigilance, and threat. They can tune immune signaling, which shapes fatigue and motivation. They can alter sleep through metabolic and inflammatory routes. These are not trivial effects. Sleep deprivation alone can amplify anxiety and reduce cognitive flexibility; chronic pain can distort attention toward threat; persistent inflammation can flatten pleasure. A gut that nudges any of these levers is a gut that nudges mood.

One concrete example comes from medicine rather than psychology: irritable bowel syndrome (IBS). IBS is not "all in the head," yet it often comes with anxiety, and gut discomfort can heighten anxious attention to bodily sensations. Many people with IBS learn, through experience, that stress can worsen symptoms, and symptoms can worsen stress. This feedback loop is a lived version of the gut—brain axis. In IBS, altered motility, visceral hypersensitivity, microbial shifts, and immune activation have all been implicated to varying degrees in different patients. No single mechanism explains all cases, but the clinical reality is that gut state and mental state can amplify one another without any moral failure on the patient's part. The body is doing what bodies do: integrating signals and adjusting behavior to perceived threat.

A historical anecdote makes the point in a different register. In the late nineteenth century, George Beard popularized the diagnosis of "neurasthenia," a condition of fatigue, anxiety, and digestive complaints attributed to modern life. The label is obsolete, and its cultural baggage is obvious, but the symptom cluster is familiar. Without modern immunology or microbiology, physicians were already noticing that mental and gut symptoms often co-occurred. Today, instead of blaming telegraphs and city living in the abstract, one can describe plausible physiological bridges: stress hormones altering gut permeability and motility; immune mediators influencing sleep and mood; microbial metabolites shaping barrier function and signaling thresholds. The old diagnosis was a blunt instrument, but it was pointing at a genuine entanglement of mind, gut, and environment.

The concept of *psychobiotics* emerged as scientists and clinicians began to ask whether certain microbes, or microbial products, might have measurable effects on mental health. The term is provocative by design. It suggests that bacteria can be, in some sense, psychoactive. Reviews in the 2010s catalogued neuroactive compounds produced by bacteria and mapped possible pathways of influence through the enteric nervous system, immune mediators, and endocrine signals. The promise is real enough to justify research, but the gaps are equally real. Human brains are shaped by experience, genetics, social context, and trauma; no microbial capsule is likely to rewrite that complexity. The more realistic hope is narrower: that in some individuals, improving gut ecology might reduce inflammatory burden, improve sleep, reduce visceral pain, and thereby soften one contributor to anxiety or low mood.

The problem, scientifically, is that the gut—brain axis is full of confounders that can fool even careful observers. People who eat more fiber often exercise more; people under chronic stress may sleep less and eat differently; medications that affect mood can affect the gut, and vice versa. Even when a microbial signal is identified, translating it into an intervention requires understanding dose, timing, and context. A metabolite that is beneficial at one concentration can be unhelpful at another. A probiotic strain that alters behavior in a mouse housed in sterile conditions may do nothing in a human whose gut ecosystem resists colonization. The strongest claims in this field are often the most modest ones: that certain microbial metabolites can change immune signaling; that vagal pathways can carry gut-derived information; that altering the gut can alter behavior in animal models under particular conditions.

This modesty is not a retreat. It is a sign of maturity. The brain is exquisitely sensitive to bodily state because bodily state predicts survival. When the immune system is activated, the brain adjusts behavior toward rest and withdrawal. When blood sugar fluctuates, attention and irritability shift. When the gut signals pain or threat, vigilance rises. Microbes participate in these signals because they are embedded in the gut's chemistry, and the gut's chemistry is embedded in the brain's priorities.

The most intimate consequence of this view is not a new supplement aisle or a new diagnostic label. It is a gentler understanding of why feeling "off" can be so hard to reason with. If the gut barrier is irritated, if immune tone is elevated, if vagal signals are biased toward alarm, the brain is not making an error when it becomes cautious and restless. It is responding to a body that is broadcasting uncertainty. In that light, care for mental life begins to look less like a contest between willpower and weakness, and more like the patient task of quieting alarms—through sleep, food, movement, social safety, and, sometimes, through medical interventions that recognize the gut as a contributor to the mind's chemical weather.

A person emerging from a period of stress often describes a particular relief: not the sudden appearance of joy, but the return of ordinary calm, as if the nervous system has stopped scanning the horizon. There is something profoundly physical about that calm. It can arrive after a good night's sleep, after a week of eating in a way that steadies the gut, after an infection resolves, or after a medication lowers inflammation. The brain is still the place where experience is felt, but the conditions that make experience feel safe are distributed through the body, including the microbial partners whose metabolites help set the background noise level of our internal world.

8.3 Engineering the Inner Garden

Once the gut is understood as a chemical marketplace rather than a simple drainpipe, the urge to intervene becomes almost irresistible. If microbes make metabolites that feed colon cells, tune immunity, and send signals that can reach the brain, then surely one can improve health by adding the "right" microbes and subtracting the "wrong" ones. The difficulty is that a gut community is not a list of species so much as a working ecology. It has supply chains, territorial boundaries, and a memory written into its chemistry. Dropping a few new organisms into that system is less like planting a rose in an empty pot than like releasing a handful of new employees into a tightly unionized workplace with its own language and schedules.

The modern public face of microbiome intervention is the probiotic: a live microorganism that, when administered in adequate amounts, confers a health benefit. That definition is already more cautious than most advertising. It does not promise that a probiotic will colonize your gut for life, or that it will "reset" anything, or that it will cure anxiety. It promises a benefit, under some conditions, at some dose. Many probiotics do have evidence for specific outcomes, such as reducing the risk of antibiotic-associated diarrhea in certain settings, or helping with particular symptoms in some individuals. Yet the most common disappointment is also the most telling: people swallow a capsule for a few weeks, feel little, and then stop. The microbes, too, mostly stop. For many products, the introduced strains pass through and fade out soon after ingestion ends.

This is not a failure of willpower. It is an ecological fact. Your resident microbes occupy niches—places defined by available nutrients, pH, oxygen, bile acids, and access to the gut lining. They are not waiting for newcomers; they are busy. They consume the carbohydrates and amino acids that arrive. They produce acids that change the local environment. They form biofilms. They stimulate immune responses that shape what can attach to the mucus layer. A newcomer must find a niche that is not already taken, or else outcompete established residents on their home turf. In most healthy adults, that is difficult.

The deeper reason is what might be called chemical mismatch. A bacterial strain is not only a genetic identity; it is a metabolic strategy. It thrives when it can reliably access the substrates it prefers and tolerate the by-products of everyone else's metabolism. A microbe adapted to ferment a particular kind of fiber may do little in a gut where that fiber rarely appears. A strain that grows well in a laboratory medium may struggle in a colon whose pH and bile acids are shaped by a different dietary pattern. Even when a probiotic survives the stomach and reaches the gut alive, it may find itself in a world that does not speak its chemical language.

Seen this way, the simplest form of microbiome engineering-adding organisms-often aims at the wrong lever. It assumes that the gut ecosystem is limited by missing species, when it may be limited by missing substrates. If you want more butyrate, you can attempt to introduce a butyrate-producing bacterium, but you can also feed the butyrate producers you already have. The second approach is frequently more realistic, because it works with the established ecology rather than against it.

This is the idea behind *prebiotics*. In everyday speech, prebiotics are sometimes described as "food for good bacteria," but scientific definitions have tightened. An international consensus statement from ISAPP (the International Scientific Association for Probiotics and Prebiotics) defines a prebiotic as a substrate that is selectively utilized by host microorganisms conferring a health benefit. The key words are *selectively utilized* and *health benefit*. It is not enough to feed microbes in general; the substrate must enrich or activate particular microbial functions in a way that measurably helps the host. Classic prebiotics include certain fermentable fibers, but the concept can extend beyond carbohydrates and beyond the gut, as long as selective utilization and benefit are demonstrated.

There is an appealing humility in that definition. It admits that the host does not fully control which microbes live inside it, but it can influence them by altering the resource landscape. A diet high in diverse plant fibers tends to increase fermentation and the production of short-chain fatty acids, shifting the chemical environment in ways that often support barrier integrity and immune balance. This does not mean fiber is a universal medicine, or that everyone should eat the same foods. Some people with sensitive guts react badly to certain fermentable carbohydrates. The point is more general: ecological engineering is often easier when it changes the conditions of life rather than attempting to replace the inhabitants.

A third category has emerged between "add microbes" and "feed microbes": *postbiotics*. The term is sometimes used loosely to mean "microbial metabolites," but ISAPP's consensus definition is more specific. A postbiotic is a preparation of inanimate microorganisms and/or their components that confers a health benefit on the host. In other words, the microbe is dead, but its cellular parts and molecular patterns still act. This seems, at first, like an odd compromise. Why take a microbe and kill it?

The answer is partly safety and partly predictability. Living microbes can evolve, exchange genes, and behave differently in different guts. They can be exquisitely sensitive to manufacturing and storage conditions. An inactivated preparation can be standardized more easily, and it cannot replicate uncontrollably. If a benefit can be achieved through microbial cell components that train the immune system, or through consistent delivery of a bacterial product embedded in a non-living preparation, then postbiotics offer a way to harness microbial biology with fewer ecological uncertainties.

Yet postbiotics also reveal a deeper truth about the microbiome: many of its benefits are chemical rather than taxonomic. The host often "cares" less about which species are present than about which metabolites are produced and which immune signals are triggered. This functional view has become increasingly influential as researchers combine microbiome sequencing with metabolomics, the measurement of small molecules. Two people can have different microbial species yet produce similar short-chain fatty acid profiles, and conversely, similar-seeming microbial communities can behave differently when diet changes. What matters to the host is the output.

A simple schematic helps distinguish two kinds of probiotic experiences that people often confuse. Some strains are *transient*: they may do something useful while they pass through, but they rarely persist. Other strains, under the right conditions, can *colonize*: they become part of the resident community for longer periods. Colonization can be desirable, but it is also harder to control.

This kind of picture clarifies why "I took probiotics and nothing changed" is not a scientific dead end. It is a question: was the strain transient, and if so, did it have a measurable effect while present? Was the intended effect even plausible without changing diet, sleep, or medication? Did the ecosystem have a niche for the strain? A person can be disappointed by a lack of colonization and still have experienced a transient functional effect, or vice versa.

If probiotics, prebiotics, and postbiotics are gentle tools, fecal microbiota transplantation (FMT) is a blunt one. It is the ecological equivalent of rewilding: introducing a whole community at once. The idea is viscerally memorable, and it has a long informal history, but modern medical FMT gained attention because of one infection in particular: recurrent *Clostridioides difficile*. After antibiotics, *C.*

difficile can bloom, causing severe diarrhea and colitis. It is difficult to treat when it recurs, and it can be life-threatening, especially in older or medically fragile patients.

There is a historical moment worth lingering on, because it captures both the ingenuity and the unease of microbiome medicine. In 1958, surgeon Ben Eiseman and colleagues reported treating patients with pseudomembranous colitis using fecal enemas. The mechanism was unknown; the microbial community itself was the therapy. It was an empirical act of ecological repair before "microbiome" became a common word. Decades later, as recurrent *C. difficile* became a recognized clinical challenge, FMT re-emerged as a remarkably effective approach in many cases, often succeeding where repeated antibiotic courses failed.

But a blunt tool brings blunt risks. Transplanting stool is not like transplanting a purified drug. Stool is a complex biological material containing live microbes, viruses, metabolites, and the traces of a donor's exposures. Even with careful donor screening, rare dangers remain. The U.S. FDA issued safety communications after serious adverse events linked to investigational FMT, including transmission concerns involving drug-resistant organisms, as well as later alerts about the potential for transmission of pathogenic *E. coli* and about the added screening needed during the emergence of SARS-CoV-2. These alerts did not invalidate the value of microbiota-based therapies; they drew a bright line under a basic principle of modern medicine: if a treatment can save lives, it must also be manufactured and monitored in a way that makes its risks legible.

This is where microbiome engineering meets regulation, and where the romance of "natural" solutions can become dangerous. When people hear "bacteria," they often imagine a category outside the usual rules of pharmacology, as if microbes were supplements rather than biological agents capable of causing infection, sepsis, or immune dysregulation. In practice, clinical use of live microorganisms demands the same seriousness as any other therapeutic. The FDA has issued guidance for early clinical trials of *live biotherapeutic products* (LBPs), focusing on chemistry, manufacturing, and control information. The phrase sounds bureaucratic, but it encodes hard-earned wisdom: if you are going to give living organisms to vulnerable patients, you must know what you are giving, how consistently you can produce it, and what contaminants you have excluded.

The field has begun to move from artisanal ecosystems toward standardized products. The FDA's approvals of microbiota-based therapies for the prevention of recurrent *C. difficile* infection marked a turning point: REBYOTA (fecal microbiota, live-jslm) was approved in November 2022, and Vowst (fecal microbiota spores, live-brpk), an orally administered product, was approved in April 2023. These approvals did not declare the microbiome "solved." They signaled that complex microbial communities can be turned into regulated biologics with defined indications, controlled manufacturing, and post-market surveillance.

The philosophical shift matters. It reframes microbiome therapy as something closer to organ transplantation than to vitamin supplementation: a deliberate alteration of a living ecosystem with consequences that can extend beyond the initial symptom. It also offers an honest model for future interventions. If the goal is to treat inflammatory bowel disease, metabolic disorders, or even certain neuropsychiatric conditions through microbial ecology, the path is unlikely to be a single over-the-counter strain marketed to everyone. It will look more like precision engineering: choosing microbial consortia (guilds of microbes that cooperate metabolically), pairing them with appropriate substrates, and monitoring functional outputs such as metabolite profiles.

Precision does not necessarily mean complexity for the patient. The sophistication can happen behind the scenes. A person may take a capsule, but the capsule could contain a designed community whose members have been selected because they collectively produce a desired set of metabolites under real gut conditions. Synthetic biology makes this prospect more than speculative. In principle, one can engineer microbes to produce specific compounds, to sense inflammatory cues, or to compete against particular pathogens. In practice, the gut is a harsh proving ground. Engineered strains must survive bile acids, avoid being swept away by peristalsis, and function amid intense competition. They must also be safe in the long term, resistant to acquiring dangerous genes, and controllable if something goes wrong.

Here again, the earlier lesson about chemical negotiation becomes central. To engineer the inner garden successfully, one must decide what is being engineered: the membership of the community, the resources that feed it, or the chemical messages it sends to the host. These are different strategies, and each has a different risk profile. Feeding resident microbes with prebiotics tends to be lower risk but can be slow and variable. Introducing live strains can be more targeted but often transient. Delivering postbiotics or other defined microbial components can be standardized but may miss the ecological context that gives a metabolite its meaning. Whole-community approaches can be powerful but demand the strictest screening and oversight.

It is tempting to treat these choices as technical details, but they have human consequences. A person with recurrent *C. difficile* is not looking for an interesting ecology lesson; they are looking for relief from debilitating diarrhea and fear of relapse. A parent searching the internet for help with a child's anxiety is not trying to become an amateur microbiologist; they are trying to quiet suffering. The microbiome field attracts hope because it offers a new kind of leverage, one that does not require changing the genome or targeting a single receptor in the brain. It offers the possibility of changing the body's baseline chemistry by changing a community.

Hope, however, has its own ecology. When a scientific idea captures the imagination, it becomes a marketplace of products, many of them premature. The most honest promise microbiome engineering can make today is also the most practical: not that it will replace psychiatry or cardiology or immunology, but that it can add a new layer of intervention—one that respects the body as an ecosystem and treats metabolites and microbial communities as real physiological actors. A well-engineered microbial therapy would not need extravagant claims. It would do something measurable: prevent a relapse, reduce an inflammatory marker, improve a barrier function assay, shorten a hospital stay, or make a symptom reliably less severe.

The inner garden is not a decorative metaphor. Gardens require attention to soil chemistry, water flow, and seasons; they reward patience and punish overconfidence. The same is true inside the gut. The most effective interventions will be the ones that treat microbes neither as magical friends nor as lurking enemies, but as partners that respond—predictably, if one is careful—to the conditions that are built for them.

Chapter 9

Blueprint for a Symbiotic Future

As we face planetary boundaries, the ancient chemical treaties of the microbial world offer a radical new operating system for human civilization. This chapter moves beyond observation to application, asking how we can harness the 'Symbiosis Engine' to redesign agriculture, medicine, and our own economic models. It argues that the future belongs not to the fittest competitor, but to the most skilled collaborator.

9.1 Engineering the Synthetic Loop

A familiar ambition in synthetic biology has been to take a single organism—a bacterium, a yeast, a plant cell—and turn it into a tiny factory. Insert a gene, delete a gene, tighten a promoter, and the cell dutifully churns out insulin, a fragrance molecule, or a biofuel precursor. This style of engineering has delivered real successes, but it also runs into a stubborn limit: cells are not empty vessels waiting to be filled with our preferred chemistry. They are crowded economies with budgets, priorities, and defensive instincts. Ask one microbe to do everything, and it often does the one thing living systems do best when overworked: it adapts, sometimes in ways that undo the engineering.

A different approach has begun to feel less like a workaround and more like a mature design philosophy. Instead of forcing a single species to carry an entire pathway, one can build a *metabolic bridge* between species: an engineered exchange in which each partner makes something the other needs, and each benefits in a way that is difficult to counterfeit. The unit of design shifts from the genome to the relationship.

This idea is older than it sounds. Long before anyone could read DNA, people learned to cultivate mixed communities whose members cooperate through chemistry. Consider the household technologies of fermentation. A sourdough starter is not a lone yeast strain heroically inflating bread by itself; it is typically a partnership between yeasts and lactic-acid bacteria, each altering the environment in ways that favor the other. The yeast produces carbon dioxide and a range of metabolites; the bacteria acidify the dough and reshape the sugar landscape; together they make a stable culture that can persist for years, passed from one kitchen to another as if it were an heirloom. The story is not only culinary. It is a quiet demonstration of an engineering principle: stability often comes from a closed loop of exchange, not from maximal self-sufficiency.

What has changed is that the loop can now be designed deliberately. In natural symbioses, partners arrive through evolution and ecological filtering. In engineered symbioses, the partnership can be assembled from parts: transporters, enzymes, sensors, and signals that determine what is shared, when it is shared, and how much is shared. The goal is not merely to place two species in the same container and hope for the best, but to make their fates entangled in a controlled, legible way.

A useful term here is *cross-feeding*. One organism releases a compound that another consumes. In nature this is everywhere: leaked amino acids, fermented acids, vitamins, hydrogen, carbon dioxide, iron-chelating molecules. Often it is accidental, a form of metabolic spillage. Yet accident is not the only option. Cross-feeding can be made into a circuit with logic: *if* you supply me with compound X, *then* I will supply you with compound Y; *if* you stop, I stop. In other words, a chemical contract.

The appeal becomes obvious when one remembers why single-organism factories fail. Many valuable molecules are hard to make because they are toxic at intermediate steps, or because they steal too much reducing power (the cell's electron budget), or because they compete with essential pathways. Cells respond by slowing growth, shutting down expression, mutating the engineered genes, or rerouting metabolism in self-protective ways. Splitting the job between two species can lower the burden on each. One partner performs an early step up to a manageable intermediate, exports it, and the second partner finishes the pathway. This is *division of labour* in its chemical sense: not a moral slogan but a way to avoid biochemical traffic jams.

There is a second, subtler advantage. When everything is inside one cell, evolution has many easy escape routes. A mutant that stops producing the desired product often grows faster, and in a large culture it can take over. In a consortium, the system can be arranged so that the very ability to grow depends on maintaining exchange. That is the promise of engineered dependency: building a loop that is self-enforcing.

Synthetic Cross-Feeding Circuit

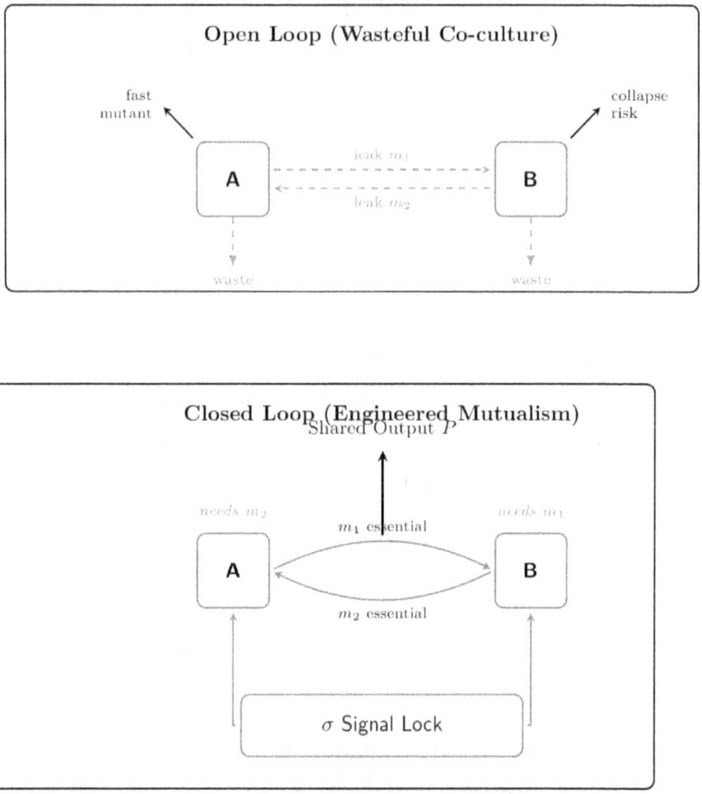

Reciprocity becomes a growth condition, not a courtesy.

The diagram sketches a contrast that matters in practice. In an open loop, two species may exchange metabolites incidentally, but nothing prevents one from outgrowing the other or from benefiting without paying. In a closed loop, exchange is turned into a growth requirement. The engineering trick is often to create an *auxotrophy*, a genetic change that makes an organism unable to synthesize a compound it normally could. If species A is made unable to produce, say, an essential amino acid, then A must obtain it from the environment—or from B. In return, B is made dependent on something A supplies. Each becomes the other's missing organ.

The word *auxotroph* can sound like jargon, but the idea is simple: remove a rung from a ladder so that the organism has to borrow that rung from its partner. Done clumsily, this is just sabotage. Done carefully, it is scaffolding for cooperation.

A concrete example makes the abstraction easier to feel. Yeast, the classic workhorse of brewing and biotechnology, can be paired with bacteria in a way that resembles a negotiated settlement. Imagine engineering a yeast strain that cannot make lysine, and a bacterial strain that cannot make adenine. On their own, each strain struggles or fails. Together, if the yeast leaks or exports adenine and the bacterium leaks or exports lysine, both can grow. Now add a second layer: regulate export so that it is conditional on receiving the partner's metabolite. Biological cells already have sensors for nutrient status and stress; synthetic biologists can wire these sensors to transporters and secretion systems. The result is a crude form of reciprocity: A shares most when A is well supplied by B. The loop is not moral, but it has a kind of logic.

Why go to this trouble? One reason is that many valuable products are easier to obtain when no single organism is forced to be both athlete and referee. Suppose the desired product is a specialty chemical that requires a harsh intermediate—a compound that damages membranes or chews up proteins. If that intermediate is kept inside one cell, the cell experiences it as self-harm. If the intermediate is exported promptly and finished by a second organism that is more tolerant, the damage can be reduced. The consortial design becomes a way to place fragile chemistry where it hurts least.

Another reason is control. A single engineered organism is a black box with too many internal degrees of freedom. In a consortium, one can distribute sensing and response. One partner can be tuned for fast detection of a pollutant; the other can be tuned for slow, thorough degradation. Or one partner can act as a buffer, taking up excess substrate that would otherwise cause runaway growth. Engineering begins to resemble electrical design: components with defined roles, linked by flows of matter.

The analogy to circuits is not merely poetic. In both electronics and biology, stability depends on feedback. A useful mutualism is often a negative-feedback system: if one partner becomes too abundant, it runs out of what the other provides, and its growth slows. If one partner becomes scarce, its metabolite becomes limiting, nudging the balance back. Yet living feedback is notoriously leaky. Molecules diffuse; cells evolve; by-products accumulate. Designing the loop means thinking in terms of rates: how quickly is a metabolite produced, how quickly is it consumed, how quickly does it escape into the bulk medium, how quickly do cheater mutants arise.

Cheaters deserve special attention because they are the natural enemy of engineered cooperation. In a cross-feeding system, a cheater might be a mutant that receives m_1 but stops paying m_2. If the environment supplies enough m_2 accidentally, or if other cells produce it generously, the cheater can thrive. In a well-designed synthetic loop, cheating should be metabolically tempting but ecologically unrewarding. That is why dependency is so central: make it hard to survive without participating.

Engineers have a repertoire of tools to enforce this, each with trade-offs. One is to couple metabolite production to essential growth processes, so that a cell cannot simply switch off sharing without harming itself. Another is to use regulated kill-switches: if a partner signal is absent, a toxin is expressed. These are powerful but delicate, because cells mutate away from lethal circuits under selection pressure. A softer method is *habitat design*. The container is part of the organism. In a well-mixed flask, a secreted metabolite becomes a public good; anyone nearby can take it. In a structured environment—a biofilm, a gel, a microfluidic device with channels—metabolites stay local, and partners can preferentially feed their immediate neighbours. Proximity becomes a policing mechanism. The loop is tightened not by coercion but by geometry.

This is one of the most counterintuitive lessons for people raised on the idea that biology is mostly genetics. Where cells live can matter as much as what genes they carry. A consortium that is unstable in a shaken tube can become stable on a surface. A partnership that fails in batch culture can persist in continuous culture, where fresh nutrients flow in and waste flows out, allowing the system to approach a steady state rather than cycling through feast and famine. These are not minor experimental details; they are control knobs.

Historically, microbiology stumbled upon this insight by accident. Sergei Winogradsky's late nineteenth-century enrichment cultures—the famous Winogradsky column—were not designed as "synthetic ecosystems," but they revealed how chemical gradients sculpt communities. Light and oxygen at the top, sulfide and darkness at the bottom: different microbes found their niche, exchanging products across millimetres. Today's synthetic biologists often recreate that principle intentionally, using micro-scale habitats to stabilize exchanges that would dissolve in a homogeneous soup. The future factory may look less like a steel fermenter full of identical cells and more like a designed landscape of interacting micro-niches.

When the loop works, it feels almost magical: two organisms that are individually fragile become jointly robust. But there is no magic, only bookkeeping. Mutualism is an accounting problem. Carbon in, energy in, nutrients in; products out; wastes out; information in the form of signals; constraints in the form of dependencies. The engineer's job is to make the ledger close.

The phrase *plug-and-play mutualism* captures the ambition. If one can map which metabolites each organism can produce and which it can consume, then one can assemble partnerships as if choosing compatible ports. This is where network mapping enters. A metabolic network is a graph of reactions and metabolites: nodes and edges, capacities and bottlenecks. With genome-scale models, one can predict which nutrient deletions will create dependencies, which exports are plausible, and where toxicity might arise. The models are imperfect, but they provide a starting point more rational than guesswork.

A particularly compelling application is waste remediation. Many pollutants are too complex to be digested efficiently by a single microbe. Breaking down a plastic additive, a pesticide residue, or a petroleum mixture often requires multiple enzymatic steps, some oxygen-dependent, some oxygen-sensitive. A consortium can split these incompatible tasks. One species might oxidize a compound to a more soluble intermediate; another might reduce it; a third might mineralize the remnants. The synthetic loop adds an extra layer: by making the partners dependent on one another, the system can be stabilized against the ecological tendency for the fastest grower to dominate and abandon the slower, essential chemistry.

Bioproduction offers an equally human incentive. Many commodities are still made from fossil feedstocks not because biology cannot make them, but because biology cannot make them *cheaply enough* and *reliably enough*. Reliability is often a question of stability over time. If a culture drifts genetically or ecologically, yields drop, and industrial reality is unforgiving. Engineered consortia aim to turn that instability into a managed variable. If one partner begins to slip, the other partner's growth is affected, changing the overall dynamics in ways that can be monitored. In principle, a consortium can be designed with built-in diagnostics: fluorescent markers, metabolite sensors, or signal outputs that report whether the exchange is functioning.

Yet humility is warranted. Natural symbioses like lichens and corals are not merely pairs of species swapping nutrients. They are layered architectures of physical structure, chemical gradients, and evolutionary history. A lichen is a miniature landscape, not a simple handshake between a fungus and an alga. A coral reef is an entire economy of recycling and retention. To mimic their stability, synthetic loops must do more than exchange two molecules. They must manage leakage, enforce reciprocity, and anticipate variability.

Variability is the quiet assassin of engineered systems. Temperature shifts change enzyme rates. A minor contaminant supplies a needed nutrient and breaks the dependency. A trace metal limitation alters which pathway is preferred. Even in the controlled world of a laboratory, a consortium can behave differently between a shaking flask and a bioreactor, between one brand of media and another. Designing the synthetic loop therefore begins to resemble designing a bridge in a windy region: one does not aim for a structure that stands only on calm days. One aims for a structure that dampens oscillations and fails gracefully.

One way to build grace into the system is to avoid absolute dependencies and instead use *tunable* dependencies. Rather than making A utterly incapable of producing m_2, one can reduce its capacity so that it can survive weakly alone but thrives with B. This can prevent catastrophic collapse if one partner temporarily dips. It is a compromise between enforceability and resilience, and it highlights a broader point: cooperation in nature is rarely an all-or-nothing contract. It is often a sliding scale, modulated by context.

Chemical signals provide another layer of tuning. Many bacteria communicate through quorum sensing, releasing small molecules whose concentration reflects local cell density. Synthetic biologists can harness such signals as the σ lock in the diagram: a chemical that must be present for the exchange machinery to turn on. If both partners contribute to σ, or if one produces σ only when fed by the other, the signal becomes a proxy for mutual presence and performance. Again, this is less a romantic language of cooperation than a control strategy: use an information channel to couple behaviours that would otherwise drift apart.

It is tempting to describe such systems as moral fables—engineered microbes learning to share—but the deeper fascination is that cooperation can be treated as a chemical phase of matter. Under some conditions, a community settles into a stable, productive configuration. Under other conditions, it flips into competition, collapse, or runaway dominance. The same ingredients, different regime. Engineers learn to ask not only "what genes do we need?" but "what regime are we creating?"

The promise of the synthetic loop, at its most practical, is that it lets human intention align with biological self-interest. If a consortium is built so that making the product is tied to staying alive, then evolution no longer automatically sabotages the design. The cells still mutate, still compete, still respond to their environment, but the space of successful adaptations is nudged toward maintaining the exchange rather than breaking it. That is not total control, and it never will be. It is something better suited to living systems: a constraint that channels freedom.

When the loop closes, waste becomes feedstock, by-product becomes currency, and survival becomes a shared project written in molecules. The remarkable thing is not that life cooperates—it always has—but that cooperation can be engineered with the same seriousness once reserved for individual genes, as if the relationship itself were a device one could build, test, and rely on.

9.2 The Diplomatic Farmer

A field of corn can look like a triumph of simplicity: one species, planted in rows, fed and watered, protected from pests, harvested with machines that move like clocks. From the air it resembles an industrial pattern stamped onto the land. Up close, the plant tells a different story. Its roots are not solitary straws but crowded trading floors. They leak sugars, amino acids, and organic acids into the soil; they release signals that invite some microbes and repel others; they adjust their chemistry when they are hungry for phosphorus or stressed by drought. The soil answers back with its own messages: hormones made by fungi, bacterial antibiotics, volatile compounds that drift through pores like a faint perfume. Every square centimetre is a conversation conducted in molecules.

Modern agriculture has been extraordinarily good at one kind of control: adding nutrients directly in the form of synthetic fertilizers. It has been less attentive to another kind of control: managing the relationships that *deliver* nutrients and protect roots under real, messy field conditions. The old dream was to make plants independent of their microbial partners by giving them everything they might want from a bag. The newer dream is closer to diplomacy. It accepts that the plant is embedded in a living network, and asks how to keep that network cooperative, diverse, and reliable.

Diplomacy begins with language. In soils, the language is chemical. Plants speak through *root exudates*—the mixture of compounds released from roots—and through more specialized *infochemicals*, molecules that carry information rather than calories. Microbes respond with their own infochemicals, and with services: nitrogen fixation, phosphorus mobilization, protection against pathogens, help with drought. No participant hears the whole conversation. Each listens for a few cues that matter to its survival, and answers in ways shaped by evolution. Agriculture changes the conversation by changing who is present, what is scarce, and what signals are worth responding to.

Synthetic nitrogen fertilizer is a good place to start, not because it is villainous, but because it is so successful that it rewired global ecology. The Haber–Bosch process, developed in the early twentieth century, made it possible to fix atmospheric nitrogen into ammonia on an industrial scale. This single technological pivot underpins much of modern food production. It also changed the incentives in the soil. If a plant can obtain abundant nitrate or ammonium without relying on microbial partners, it may invest less in recruiting them. If the soil receives repeated pulses of nitrogen, microbial communities shift toward those that thrive under high nutrient availability. The economic metaphor becomes almost literal: flood the market with a cheap commodity, and long-standing trading relationships weaken.

Phosphorus, too, enters the story. Unlike nitrogen, phosphorus has no atmospheric reservoir that can be industrially captured; it is mined from rock and shipped around the world. In many soils phosphorus is present but locked up in insoluble forms, chemically bound to minerals. Plants often rely on mycorrhizal fungi—the thin, thread-like fungi that colonize roots—to scavenge phosphorus and other nutrients in exchange for carbon. When fields receive large amounts of readily available phosphorus fertilizer, the plant's need for fungal help can diminish. The relationship becomes optional, and optional relationships are vulnerable. The fungi may still colonize roots, but the balance of costs and benefits changes. Under some conditions the plant pays carbon but receives little that it could not have obtained alone.

Calling this a conversation also helps highlight what industrial simplification removes. A diverse plant community maintains a diverse chemical vocabulary: different root architectures, different exudates, different seasonal rhythms. Monocultures reduce that vocabulary. They can create a repetitive signal landscape that favors a narrower set of microbes, including pathogens that specialize on that crop. Herbicides and broad-spectrum pesticides can further narrow the cast of characters. Deep tillage disturbs fungal networks and breaks up soil structure, turning a complex habitat of pores and aggregates into something closer to a shaken flask. Even when the crop looks healthy above ground, the underground social order may be depleted.

Yet the soil is not a passive victim. It is an arena of negotiation, and plants are not naive hosts. Some of the most striking evidence comes from the legume–rhizobium mutualism, the partnership that allows peas, beans, and soybeans to obtain nitrogen through bacterial fixation. Rhizobial bacteria live in nodules on the roots and convert atmospheric N_2 into ammonia, a form the plant can use. This is energetically expensive for the bacteria, and it raises an obvious problem: why not cheat? Why not occupy nodules, take shelter and food, and fix little or no nitrogen?

A key discovery by E. Toby Kiers and colleagues in 2003 made the plant's answer hard to ignore. Working with soybeans, they showed that plants can impose *sanctions* on rhizobia that fail to fix nitrogen. In nodules containing ineffective bacteria, the plant reduced the bacteria's reproductive success dramatically, plausibly by limiting the oxygen supply the bacteria need for metabolism. Oxygen is a double-edged tool in nodules: rhizobia require it to generate energy, but too much oxygen damages nitrogenase, the enzyme that performs nitrogen fixation. Plants already manage oxygen with exquisite care. The insight was that this oxygen valve can also function as enforcement. The plant is not merely providing a home; it is monitoring performance and adjusting rewards.

Sanctions matter because they show what stable cooperation often requires: the ability to measure, and the ability to respond conditionally. A diplomatic farmer learns from this botanical lesson. The goal is not only to introduce "beneficial microbes" but to create conditions where the plant can reward the helpful and restrict the unhelpful, and where microbes that play well with others can persist across seasons.

Mycorrhizal fungi offer a second, more complicated arena. Arbuscular mycorrhizal fungi (AMF) penetrate root cells and form tree-like structures called arbuscules, through which nutrients and carbon are exchanged. If rhizobium nodules look like specialized organs, AMF look like extensions of the root itself. Their hyphae explore soil pores too small for roots, delivering phosphorus, nitrogen, and micronutrients, and sometimes improving water relations. They can also reduce nutrient loss by helping retain nitrogen and phosphorus in the soil system rather than letting them wash away.

Farmers and agronomists have long wanted to harness AMF more deliberately. One approach is inoculation: adding fungal spores or colonized root fragments to soil or seedlings. Meta-analyses of many experiments suggest average benefits can be substantial for biomass and nutrient uptake. At the same time, those averages hide a frustrating truth that every field agronomist knows: context dominates. In laboratory pots with sterilized soil, an inoculant can look miraculous. In real fields with established microbial communities, variable weather, and complex soil chemistry, the effect can shrink, disappear, or even flip sign.

A diplomatic view makes this less mysterious. In a real field, the introduced fungus is not arriving to an empty stage. It must compete with local fungi and bacteria, survive predation, and find hosts willing to pay carbon. The soil may already have abundant mycorrhizal inoculum. If phosphorus is high, the plant's incentive to invest in the symbiosis may be low, reducing colonization. If the soil has been tilled heavily, fungal networks are repeatedly disrupted, and the ecology favors fast-recovering generalists rather than slow-building mutualists. The inoculant is not simply a product; it is an immigrant entering a crowded city with its own politics.

The chemical conversation of the rhizosphere—the thin zone of soil influenced by roots—is also shaped by how plants allocate carbon. Carbon is the currency plants can spend, and their spending changes with light, temperature, water, and nutrient status. A drought-stressed plant may close stomata and reduce photosynthesis, shrinking the carbon budget available for microbial partners. In that moment, microbes may behave differently: some will become more competitive, others more cooperative, some will switch from mutualistic roles to opportunistic ones. This is one reason the same microbial inoculant can behave differently from one season to another. The partner is not merely the microbe; it is the plant under specific conditions.

There is a temptation, especially in the commercialization of agricultural "biologicals," to talk as if a beneficial microbe is like a vitamin pill: add it, and health improves. Diplomacy offers a harder but more realistic framing: a beneficial microbe is like a skilled worker. Whether it improves the farm depends on whether the job exists, whether it has the right tools, whether it is paid, and whether the rest of the community allows it to function.

This way of thinking changes what interventions look like. Instead of asking only, "Which microbe should be added?" one asks, "What habitat makes cooperation likely?" Soil structure becomes a primary variable. Reduced tillage can preserve fungal hyphal networks and soil aggregates, which create microhabitats where beneficial microbes can persist. Cover crops add continuous root presence, maintaining a steady trickle of exudates and preventing the rhizosphere from collapsing between cash crops. Diverse rotations shift the chemical vocabulary of the soil across time, discouraging specialized pathogens and broadening the mutualist pool. Organic amendments can feed microbial communities, but they can also shift competition in ways that favor decomposers over symbionts; again, the outcome depends on composition and timing.

Even fertilizer management can be reframed. The question is not simply how to maximize yield this year, but how to balance inputs with the maintenance of symbiotic capacity. Over-application of nitrogen can suppress biological nitrogen fixation and contribute to nitrate leaching; over-application of phosphorus can reduce reliance on mycorrhizas and increase phosphorus runoff, a driver of eutrophication in rivers and lakes. A diplomatic strategy is sometimes to apply less, but also to apply differently: split applications, targeted placement near roots, or timing that avoids flooding the soil with nutrients when plants could otherwise recruit microbial help. The aim is not to romanticize low-input agriculture, but to use inputs without silencing the soil's own services.

The stakes of this conversation reach beyond the farm gate. Nutrient losses from agriculture are among the most visible ways that broken symbioses become public problems. Nitrogen and phosphorus that escape fields do not vanish; they move into waterways, fueling algal blooms and oxygen-depleted "dead zones." From a systems perspective, this is a failure of retention. Healthy symbiotic networks, especially those involving mycorrhizal fungi, can act as retention infrastructure: capturing nutrients, cycling them, and slowing their escape. It is not a perfect seal, but it can be a meaningful difference between a landscape that holds fertility and a landscape that leaks it.

There is also a human story embedded here, one that echoes the earlier history of fermentation. Before synthetic fertilizers, farmers depended heavily on rotations with legumes, manures, and fallows to maintain fertility. These were not always efficient or pleasant, and they could not easily feed today's population at today's expectations. The fertilizer revolution was, in many regions, a liberation from the tight constraints of local nutrient cycling. The diplomatic approach does not demand a return to pre-industrial scarcity. It tries to keep the gains of modern agronomy while rebuilding the biological infrastructure that makes those gains sustainable.

One can see the beginnings of this shift in the growing interest in *diazotroph* inoculants for non-legume crops. Diazotrophs are nitrogen-fixing microbes, and the idea of helping maize or wheat obtain some nitrogen biologically is alluring. Field trials and commercial products suggest that under some conditions such inoculants can boost yield and reduce the need for fertilizer. Under other conditions, they do little. The diplomatic interpretation is again straightforward: nitrogen fixation is expensive, and microbes will not do it reliably unless conditions make it worth their while and the plant provides a hospitable niche. The challenge is not only identifying strains with the right genes, but creating a consistent ecological role for them in diverse soils.

Consistency is where agriculture differs sharply from many industrial biotechnologies. A bioreactor can be sterilized. A field cannot. A field is not a vessel but an ecosystem exposed to weather, wildlife, and history. This is why soil partnerships can feel like a frustrating frontier: full of promise, stubbornly variable. But diplomacy is precisely the art developed for dealing with autonomous partners in variable conditions. It works by combining incentives, enforcement, and communication.

In plant terms, incentives are carbon flows and niche opportunities; enforcement can be sanctions, selective feeding, or immune responses; communication is the chemical vocabulary of exudates and microbial signals. In farmer terms, incentives are habitat and management choices; enforcement is sometimes as simple as avoiding practices that reward freeloaders (for instance, chronic nutrient flooding that makes microbial services irrelevant); communication includes seed coatings, amendments, and crop choices that shape root chemistry. None of this is as simple as "add microbe, get benefit," but it is closer to how the soil actually behaves.

There is a further, more personal reason this matters. Soil is often described as dirt, as background, as something to be moved and leveled and corrected. But anyone who has watched a field after a heavy rain knows the difference between living structure and exhausted powder. One absorbs water, smells earthy, resists erosion, and supports crops through dry spells. The other crusts, runs off, and demands ever more inputs to stay productive. Those differences are not just physical; they are social, in the microbial sense. They reflect whether the chemical conversation is rich enough for cooperation to emerge.

A diplomatic farmer, then, is not someone who abandons modern tools. It is someone who treats the rhizosphere as a partner rather than an inert medium, and who learns to listen for the consequences of management in the language of nutrients retained, roots protected, soils aggregated, and yields stabilized across bad years as well as good ones. The reward is a kind of resilience that cannot be poured from a bag: a field that carries within it a memory of cooperation, stored in the living networks that make fertility less fragile and agriculture less lonely.

9.3 The Logic of Long-Term Exchange

Cooperation in nature is rarely a sentimental truce. It is a technology for surviving time. When two partners keep trading over generations, they do so in a world that contains droughts, parasites, freeloaders, and sudden windfalls. The surprising thing is not that cooperation exists, but that it persists when temptation is constant. The chemistry of cooperation is therefore inseparable from a second theme: the logic of long-term exchange. It is the logic that turns a one-off bargain into a durable relationship.

Human societies have their own long-term exchanges: contracts, markets, laws, reputations, norms. These institutions are often treated as inventions that lift us out of nature. Yet when one looks closely, they rhyme with biological solutions to the same problem: how to keep benefits flowing when each partner could gain, in the short run, by taking without giving. A farmer wants reliable rainfall; a city wants stable food supply; a company wants predictable inputs; a patient wants a medicine that works the same way every time. Time punishes systems that only win today.

The most basic difficulty is the one that economists call a *public good*: something beneficial that is hard to exclude others from. In microbial communities, public goods are often secreted molecules—enzymes that digest food outside the cell, iron-scavenging compounds that capture scarce metals, extracellular matrices that hold a biofilm together. These can make the whole community more productive. They also invite exploitation. A mutant that stops producing the public good can often grow faster, because it saves energy while still enjoying the shared benefit. In a well-mixed environment, such cheats can spread quickly, dragging the entire community down.

The story is not limited to microbes. It is the same structural problem faced by fisheries, groundwater, clean air, and climate stability: benefits are shared; costs are individualized; temptation is ever-present. The language changes—carbon emissions instead of extracellular enzymes—but the mathematics of incentives is familiar. Nature's solutions are not moral lectures. They are mechanisms: ways of making exploitation harder, less profitable, or more detectable.

One mechanism is *structure*. In a biofilm, cells live in a matrix they secrete, often attached to a surface. Structure changes who benefits from what. A cell that produces a useful enzyme is more likely to feed its close neighbours than distant strangers. If the producing cell tends to be surrounded by its own descendants (because cell division keeps kin nearby), then the producer indirectly benefits from helping relatives. This is not altruism so much as geometry plus heredity. Cooperation becomes stable not because everyone behaves well, but because the benefits are channelled toward those who share the cost.

Human analogues of structure are everywhere. Local irrigation systems often work best when farmers share a canal and can see who is maintaining it. Neighbourhood-level waste collection can be more reliable than anonymous dumping grounds. The details differ, but the principle holds: when the effects of an action remain local and visible, cooperation has a foothold.

A second mechanism is *constraints.* In engineered microbial consortia, a constraint can be a designed dependency: one partner cannot grow unless it receives a metabolite from the other. In natural systems, constraints may be physiological. A coral polyp, for example, relies on photosynthetic symbionts for much of its energy, but the symbionts also rely on the coral for nutrients and a protected habitat. The partners are bound by the fact that neither thrives, in that ecological niche, alone. Constraints do not eliminate conflict. They simply narrow the range of viable selfish moves.

The human temptation is to treat constraints as punishment, as something imposed from above. In biology, constraints are more subtle. They can be built into the arrangement of exchange. A plant can route carbon through specific tissues; a host can limit access to resources; a symbiont can evolve to depend on host-provided compounds. Constraints can be a form of design that reduces the need for constant policing.

Policing, however, is the third major mechanism, and it deserves its own name: *enforcement*. The legume sanctions against non-fixing rhizobia remain one of the clearest examples because they show enforcement operating in real time. The plant has a way to assess performance (nitrogen fixation) and a way to alter payoffs (oxygen supply, and thereby bacterial reproduction). The key is conditionality: reward is not granted once and forever; it is contingent on continued contribution.

Human systems have developed enforcement mechanisms for long-term exchange: inspections, audits, warranties, legal penalties, reputation systems. These tools work imperfectly, and they can be abused, but their function is recognizably biological: align incentives by making cheating costly.

The interesting question is not whether enforcement is necessary, but what kind of enforcement works when partners are complex, semi-autonomous entities rather than simple machines. In microbial communities, enforcement is often chemical and local. In human societies, it is institutional and often remote. This distance can be dangerous. When the people harmed by a broken exchange are far away in space or time, enforcement weakens. The atmosphere is the ultimate well-mixed flask.

A historical anecdote from economic life can illuminate why long-term exchange needs more than goodwill. During the nineteenth century, many cities struggled with adulterated food: milk diluted with water, flour whitened with chalk, spices bulked out with fillers. Consumers could not easily detect cheating. The market alone did not reliably punish it. Over time, societies built institutions of verification: standardized measures, food inspection, liability laws, and later, chemical assays. This was not a triumph over nature but an application of nature's logic: the more difficult it is to monitor performance, the more likely exploitation becomes, and the more a system needs mechanisms that make hidden cheating visible.

A similar verification problem appears today in efforts to deploy living microbes as products, whether in agriculture or medicine. A microbial inoculant is not a static chemical. It is a population that can change during manufacturing, storage, and after release. Ensuring identity and function becomes a question of measurement: which strains are present, at what abundance, and with what activity. Without reliable measurement, long-term exchange between producer and user becomes fragile, because trust cannot be anchored in evidence. One can call this regulation, but it is also diplomacy: a framework that lets partners cooperate at scale without needing personal familiarity.

There is, however, a trap in borrowing human economic metaphors too casually. It is tempting to describe mutualisms as markets where partners barter carbon for nutrients, setting prices and shopping for the best deal. The *biological market* metaphor has been useful, especially for emphasizing choice and competition among partners. Yet in some systems, particularly mycorrhizal symbioses, evidence increasingly suggests that exchange may not behave like a neat price-coupled barter. Carbon flow from plant to fungus and nutrient flow from fungus to plant can be shaped by source—sink dynamics: where carbon is produced and stored, where nutrients are concentrated or scarce, how fast they can be transported, and what physical constraints limit transfer.

This distinction matters because it changes what one should design or regulate. If exchange is price-like, one might focus on making markets more efficient: more partner choice, better information, more competition. If exchange is dominated by sinks, stoichiometry, and constraints, then "market design" is the wrong mental tool. The relevant levers become habitat structure, nutrient availability, and enforcement mechanisms that prevent exploitation under those constraints. A mechanism-first view is less romantic but often more predictive.

Stoichiometry is an unglamorous word for a powerful idea: living bodies must maintain certain ratios of elements. Carbon, nitrogen, phosphorus, iron, and other nutrients cannot substitute for one another. A plant can be awash in carbon and still be limited by phosphorus. A fungus can have access to nitrogen and still be limited by energy. In such cases, the "price" of a nutrient may not be negotiated so much as dictated by physical shortage. Partners may exchange not because they have optimized a bargain, but because their constraints lock them into mutual dependence.

This has a sobering implication for human sustainability. Many environmental problems are treated as if they were purely economic coordination failures: set the right price, and the system will self-correct. Pricing can help, but biology suggests that long-term exchange often fails not only because incentives are misaligned, but because the physical system leaks, concentrates, or breaks feedback loops. If nutrients wash out of a field into a river, that is not just a moral failure; it is a structural one. If carbon dioxide accumulates in the atmosphere, it is not because the atmosphere is negotiating poorly; it is because the sink is slow and the mixing is global. Durable cooperation often requires changing the physical pathways through which benefits and harms travel.

Nature also warns against a simplistic view of "win—win." Many symbioses persist even when the benefits are uneven or fluctuate across conditions. Corals and their algal symbionts can be extraordinarily productive in stable conditions, yet the relationship can become fragile under heat stress, leading to bleaching. In some circumstances, partners that are usually helpful can become harmful. The long-term logic is not that every interaction is permanently mutualistic, but that systems can tolerate tension when they have mechanisms for recovery, partner replacement, or fallback strategies.

This is where the idea of *tradeoffs* becomes central. In both engineered consortia and natural ecosystems, a design that maximizes output under ideal conditions may be less stable under stress. A microbial community tuned for high production may be vulnerable to cheaters. A farm system optimized for maximum yield may be vulnerable to drought or disease if it has stripped away microbial and plant diversity. A city optimized for efficiency may be brittle when supply chains are disrupted. Long-term exchange demands not the elimination of tradeoffs but their explicit management.

One can think of stability as an emergent property built from three kinds of ingredients: constraints that make partners interdependent, structure that localizes benefits and makes contributions harder to steal, and enforcement that detects and penalizes persistent non-cooperation. These ingredients appear in different guises across biology and society. None is sufficient alone. Constraints without structure can still be exploited if public goods leak. Structure without enforcement can still fail if cheats find a way around local feedbacks. Enforcement without legitimacy can provoke resistance and instability, whether in microbial populations evolving around a kill-switch or in human communities evading unfair rules.

A concrete illustration, close to everyday life, is the way cities manage wastewater. For much of history, waste was simply dumped into rivers or streets. The immediate benefits of disposal were local; the harms accumulated downstream. Over time, societies built sewage systems and treatment plants that close loops by turning waste into something less harmful, sometimes even into fertilizer or biogas. This is a form of engineered symbiosis at the scale of infrastructure: a designed pathway that prevents leakage and makes the long-term costs of short-term convenience harder to externalize. It works only with measurement (water quality monitoring), structure (pipes and plants), and enforcement (regulations and penalties). It is not an accident that these are the same categories that stabilize microbial and plant mutualisms.

The deeper lesson is that cooperation is not a single virtue. It is a continuous performance under changing conditions, and it requires feedback. In biological systems, feedback is often chemical: a host changes nutrient supply; a symbiont changes metabolite release; signals ramp up or shut down pathways. In human systems, feedback is informational and institutional: monitoring, reporting, and adaptive rules. When feedback fails, relationships drift into exploitation or collapse.

There is also a psychological discomfort in acknowledging enforcement as a component of cooperation. People often prefer stories in which harmony emerges spontaneously from mutual interest. Biology is less squeamish. It shows that stable cooperation frequently relies on the ability to say no: no oxygen to a non-fixing rhizobium, no access to a niche for a pathogen, no free ride in a structured biofilm where producers feed their kin. This does not make nature punitive. It makes nature realistic about incentives.

Realism can be hopeful. It implies that sustainable human systems do not require angels. They require design that assumes temptation and builds around it. The most robust relationships are not those in which cheating never occurs, but those in which cheating does not pay for long, and in which cooperation is easier than sabotage.

That is why the earlier image of a "synthetic loop" matters beyond biotechnology. A loop is a path that brings consequences back to their source. When a factory's waste becomes another process's feedstock, the loop tightens. When a farm retains nutrients rather than leaking them, the loop tightens. When a company is accountable for the lifecycle of its products, the loop tightens. Loops do not abolish conflict, but they transform it. They replace the fantasy of costless extraction with the practical discipline of reciprocity.

In living systems, the final arbiter is survival over time. Partnerships that cannot manage their internal tensions do not persist, no matter how productive they look in the short run. Human systems have a choice that microbes do not: to learn from this logic without waiting for collapse to teach it. The chemistry of cooperation is a reminder that durability is built, molecule by molecule and rule by rule, by making it easier to keep promises than to break them, and by arranging the world so that what is taken must eventually be returned.

www.ingramcontent.com/pod-product-compliance
Lightning Source LLC
LaVergne TN
LVHW022206200926
848481LV00006B/1863